D0163012

Estimating
Air Conditioning
Systems

OTHER RESTON BOOKS BY BILLY LANGLEY

Air Conditioning and Refrigeration Troubleshooting Handbook
Basic Refrigeration
Comfort Heating
Heat Pump Technology
Refrigeration and Air Conditioning

Estimating
Air Conditioning
Systems

BILLY C. LANGLEY

RESTON PUBLISHING COMPANY, INC.
A Prentice-Hall Company
RESTON, VIRGINIA

Library of Congress Cataloging in Publication Data

Langley, Billy C.
  Estimating air conditioning systems.

  Includes index.
  1. Air conditioning—Estimates. I. Title.
TH7687.5.L365   1983   697.9'3   82-21630
ISBN 0-8359-1790-8

Editorial/production supervision and interior design
by Barbara J. Gardetto

© 1983 by
Reston Publishing Company, Inc.
A Prentice-Hall Company
Reston, Virginia 22090

*All rights reserved. No part of this book may be
reproduced in any way, or by any means, without
permission in writing from the publisher.*

10   9   8   7   6   5   4   3   2

Printed in the United States of America

Contents

Preface

Estimating Air Conditioning Systems is designed to provide the theory and practice of sizing a complete air-conditioning system, designing the ductwork, and cost estimation. It is intended to be used as a curriculum guide, a textbook, or a course for independent study.

The text consists of nine chapters, each of which covers a specific area of system estimating. Each chapter begins with an introduction to that particular phase of study and advances through recognized estimating procedures. A summary and questions that cover the minimum material with which a reader should be familiar conclude the first eight chapters. Chapter 9 is a guided exercise in complete system estimating.

Billy C. Langley

Air Conditioning and Psychrometrics

1

Most people have little appreciation of the basic principles of air conditioning, probably because the public in general only became conscious of air conditioning in about 1920. It was at this time that large-scale use of air conditioning began on trains and in theaters. It was these systems that first exposed large numbers of people to the comfort and advantages of summer cooling and at the same time caused the erroneous impression that air conditioning is synonomous with cooling.

INTRODUCTION

The term *air conditioning* has for many years been misunderstood. Air conditioning involves treating all the properties of air. Total air conditioning involves circulation, cooling, heating, humidifying, de-humidifying, and cleaning the air. A working knowledge of these processes must be understood before satisfactory estimation of air conditioning systems can be accomplished.

The heating process involves air circulation, heating, humidification, and cleaning. The cooling process involves air circulation, cooling, dehumidifying, and cleaning.

DEFINITION

The accepted definition of the term *air conditioning* is: the simultaneous mechanical control of temperature, humidity, air purity, and air motion. Unless all these conditions are controlled, the term *air conditioning* cannot be applied properly to any system or equipment. It should be noted that the control of humidity can mean either humidifying or dehumidifying. Thus an industrial system that provides an indoor condition of 150°F (66°C) dry bulb at 75% relative humidity can just as well be called air conditioning as a system designed to provide indoor conditions of 80°F (26.7°C) dry bulb and 50% relative humidity. Similarly, a system that only cools a space, without regard to the relative humidity or air purity or motion, cannot properly be called a true air conditioning system. An air conditioning system can maintain any atmospheric condition regardless of variations in the outdoor atmosphere.

HUMAN COMFORT

The two primary reasons for using air conditioning are (1) to improve the control of an industrial process, and (2) to maintain human comfort. The conditions to be maintained in an industrial process are dictated by the very nature of the process or the materials being handled. In a comfort system, however, the conditions to be maintained are determined by the requirements of the human body. Therefore, an understanding of the essential body functions is basic to an understanding of air conditioning.

The comfort of a human body is dependent on how fast the body loses heat. The human body might be compared to a heating unit that uses food as its fuel. Food is composed of carbon and hydrogen. The energy contained in the fuel, food in our case, is released by oxidation. The oxygen used in the process comes from the air, and the principal products of combustion are carbon dioxide and water vapor. Doctors call this process *metabolism*.

The human body is basically a constant-temperature machine. The internal temperature of the body is 98.6°F (37°C), which is maintained by a delicate temperature-regulating mechanism. Because the body always produces more heat than it needs, heat rejection is a constant process. The main object of air conditioning is to assist the body in controlling the cooling rate. This is true for both the heating and cooling seasons. In summer, the job is to increase the cooling rate; in winter, it is to decrease the cooling rate.

In the air conditioning process, there are three ways in which the body gives off heat: (1) convection, (2) radiation, and (3) evaporation. In most cases the body uses all three methods at the same time.

■ Convection

In the convection process, the air close to the body becomes warmer than the air farther away from the skin. Because the warm air is lighter than the cool air, the warm air rises. This warm air is replaced by the cooler air and the cooling by convection is a continuous process. As this air becomes warm, it also floats upward. Even though the deep body temperature remains at 98.6°F (37°C), the human skin temperature will vary. The skin temperature may vary from 40 to 105°F (4.4 to 40°C) in relation to the temperature, humidity, and velocity of the surrounding air. If the temperature of the surrounding air drops, the temperature of the skin will also drop.

■ Radiation

Heat radiates directly from the body to any cooler surface just as the rays of the sun travel through space and warm the surface of the earth. Heat may flow from the skin to any surface or object that is cooler than the body. This process is independent of the convection process. The temperature of the air between the person and the cooler surface has no effect on the radiation process. The same principle applies when a person is warmed by a camp fire. The side next to the fire gets warm while the other side is cool. The air temperature between the person and the fire is the same as the air temperature on the person's other side.

■ Evaporation

Evaporative heat regulation is the body process that maintains life outside an air conditioned space. In this process moisture, or perspiration, is given off through the pores of the skin. When this moisture evaporates, it absorbs heat from the body and cools it. The effect of evaporation can be felt more easily when alcohol is put on the skin because the alcohol vaporizes more readily and absorbs heat faster. This evaporation turns the moisture into low-pressure steam or vapor and is a continuous process. When drops of sweat appear on the skin, it means that the body is producing more heat than it can reject at the normal rate.

CONDITIONS THAT AFFECT BODY COMFORT

There appears to be no set rule as to the best conditions for all people. In the same atmospheric conditions, a young person may be slightly warm whereas an elderly person may be cool.

The three conditions that affect the ability of the body to give off heat are (1) temperature, (2) relative humidity, and (3) air motion. A change in any one of these conditions will either speed up or slow down the cooling process.

■ Air Temperature

Air at a temperature lower than the skin will speed up the convection process. The cooler the air, the more heat the body will lose through convection. Heat always flows from a warm place to a cooler place. The greater the temperature difference, the faster the heat will flow. If this difference in temperature is too great, the body will lose heat more rapidly than it should and will become uncomfortable.

If the air temperature is higher than the skin temperature, the convection process will be reversed. The body heat will be increased. It can be seen that the air temperature has a very important effect on human comfort. Experience shows that air temperatures ranging from 72 to 80°F (22.2 to 26.7°C) feel comfortable to most people.

The temperature of any surrounding surfaces is also important because this temperature affects the rate of radiation from the body. The lower the surface temperature, the more heat is given off by the body through radiation. As the temperature difference between the surface and the body is decreased, the rate of radiation is decreased. The radiation process will be reversed if the surrounding surface temperature is higher than the body temperature. When this happens, the body must give off more heat through the convection and evaporation processes.

■ Relative Humidity

Relative humidity regulates the amount of heat that the body can reject through radiation. Relative humidity is a measure of the amount of moisture in the air. It is an indication of the ability of the air to absorb moisture. Relative humidity is basic to the air conditioning process.

As an example, let us consider 1 cu ft (0.028 m^3) of air at 70°F (21.1°C), which contains 4 grains (0.26 g) of water vapor (see Figure 1-1). A grain is a very small amount of water. Actually, it takes 7000 grains (453.69 g) to make 1 lb.

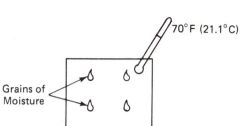

FIGURE 1-1
One cubic foot of air at 70°F (21.1°C) with 4 grains (0.26g) of water vapor.

The relative humidity can be determined in the following manner. There are only 4 grains (0.26 g) of water vapor in the 1 cu ft (0.028 m³) of air. If that cubic foot of air held all the moisture it could possibly hold at that temperature, it would actually hold 8 grains (0.52 g) of water vapor and would be said to be saturated (see Figure 1-2).

To find the relative humidity, divide the moisture actually present in the air by the amount that the air could hold at a saturated condition at the same temperature (see Figure 1-3). This process tells us that the relative humidity is 50%. Relative humidity is an indication of the amount of moisture in the air compared to the amount of moisture that could be present at that temperature. The relative humidity will change with a change in temperature.

For example, the temperature of the air is raised to 92°F (33.3°C) without adding any moisture. When the humidity tables are checked, we can see that 1 cu ft (0.028 m³) of air at 92°F (33.3°C) will hold 16 grains (1.04 g) of water vapor when saturated. The relative humidity in this example is 4 grains (0.26 g) divided by 16 grains, or 25% (see Figure 1-4).

If air that surrounds the body has a low relative humidity, the body will give off more heat through evaporation. If the air surrounding the body has a high relative humidity, the body will give off less heat through the evaporation process. A conditioned air temperature at 80°F (26.7°C) and 50% relative humidity will be reasonably comfortable.

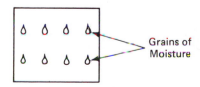

FIGURE 1-2
Cubic foot of saturated air.

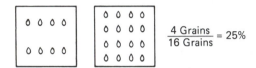

FIGURE 1-3
Determination of relative humidity.

$$\frac{4\text{ Grains}}{16\text{ Grains}} = 25\%$$

FIGURE 1-4
Relative humidity at a higher temperature.

■ Air Movement

An increase in the rate of evaporation of perspiration from the body is largely a result of air movement. Evaporation is dependent on the ability of air to absorb moisture. As the air moves across the skin, the moisture-laden air is replaced by drier air, which allows more moisture to evaporate from the skin.

If the air is allowed to remain still (static), the air next to the skin will absorb moisture until the saturation point is reached. As the saturation point is reached, the evaporation process is slowed down. The moisture will evaporate more slowly and will eventually stop when the saturation point is reached and the person will feel uncomfortable.

The movement of air also speeds up the convection process. This is possible because the warm air next to the skin is replaced by cooler air and heat is given up from the body to the air.

Air movement also removes heat from other substances, such as walls, ceilings, and other objects surrounding the body, thus tending to speed up the radiation process. It must be remembered that air motion is one of the conditions that affects the comfort of human beings.

AIR CONDITIONING TERMINOLOGY

Before a study of air conditioning can be undertaken, an understanding of the terms used is necessary. The following definitions explain the most common terms.

Dry Air.   Dry air is air that contains no water vapor. In nature, air contains some water vapor.

Absolute Humidity.   Absolute humidity is the actual quantity or weight of moisture present in the air in the form of water vapor. It is expressed as the weight of water vapor in grains per pound of air. There are 7000 grains of moisture to 1 lb (0.4536 g) of moisture. A rise or fall in temperature without a change in the quantity of moisture in the air does not affect the absolute humidity.

Relative Humidity.   Relative humidity (RH) is the ratio between the moisture content of a given quantity of air at any temperature and pressure compared to the maximum amount of moisture which that same quantity of air is capable of holding at the same temperature and pressure. Relative humidity is expressed as a percentage. Note particularly the difference between relative humidity and absolute humidity. The relative humidity can be found by dividing the amount of moisture actually present by the maximum moisture-holding capacity of the air at that same temperature and pressure.

For example, if 1 cu ft (0.028 m³) of air contains 4 grains (0.26 g) of moisture and is capable of holding 8 grains (0.52 g) of moisture, what is the relative humidity?

$$\text{RH} = \frac{\text{absolute humidity}}{\text{maximum amount air can hold}} = \frac{4}{8} = 50\%$$

A rise in air temperature will cause a decrease in the ratio or relative humidity. A fall in air temperature will cause a rise in the ratio or relative humidity, unless the saturation point (100% RH) has been reached. Beyond this point, condensation will take place as the temperature falls. When air contains all the moisture it can hold at any given temperature and pressure, it is said to be saturated. When air is saturated, the relative humidity is 100%.

Humidification.   The process of adding moisture to the air is called humidification. This process is usually accomplished with a humidifier.

Dehumidification.   The process of removing moisture from the air is called dehumidification. This moisture removal is accomplished by lowering the air temperature below the condensation point. This process is usually accomplished by use of a refrigeration evaporator coil located in the airstream.

Saturated Air.   Saturated air is air that contains all the water vapor it is capable of holding at that particular pressure and temperature. When air is saturated the dry bulb, the wet bulb, and the dew point temperatures are the same.

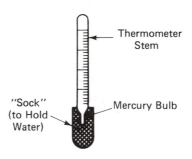

FIGURE 1-5
Wet bulb thermometer.

Dry Bulb Temperature.   The temperature of the air as indicated by an ordinary thermometer is known as the dry bulb temperature and is a measure of the sensible heat content of the air.

Wet Bulb Temperature.   The wet bulb temperature is the temperature of the air as measured by an ordinary thermometer with the bulb covered with a wet cloth or gauze (see Figure 1-5). The temperature is taken after the thermometer has been exposed to a rapidly moving airstream. The temperature indicated on a wet bulb thermometer will be depressed lower than a dry bulb thermometer reading because of the evaporation of the moisture in the wick. The heat necessary for the evaporation of the water from the wick is removed from the air while the air is passing over the thermometer. The thermometer has been exposed to the airstream for a sufficient length of time when the temperature falls to a point of equilibrium which is determined by the rate of moisture evaporation from the wick and the quantity of sensible heat in the air.

The point of equilibrium is the temperature at which the rate of sensible heat transfer from the air to the water on the wet bulb is equal to the rate at which heat is transferred from the wet bulb to the air by evaporation of the water. For any given dry bulb temperature and any given relative humidity, the point of equilibrium will be the same. Since the rate of evaporation from the wick is determined by the quantity of moisture in the air, as well as the sensible heat, it can be seen that the wet bulb temperature is indicative of the total heat contained in the air.

Wet Bulb Depression.   The difference between the dry bulb temperature and the wet bulb temperature for any given condition is called the wet bulb depression, except at the saturation point, where the two temperatures are the same.

Dew Point Temperature. The dew point temperature is the temperature at which water vapor will start to condense out of the air. The quantity of water vapor in the air is always the same at any given dew point temperature. Therefore, the quantity of moisture in the air can be measured by the dew point temperature. When air is at the dew point temperature, it is holding all the moisture it can hold at that temperature. As long as there is no removal or addition of moisture, the dew point temperature will remain constant. No latent heat is given to or released from a mixture of air and water vapor unless some moisture is added to or removed from the air.

Saturation Temperature. When air is at the dew point temperature, it is also at the saturation temperature and is considered to be saturated with moisture. The relative humidity is 100% at the saturation temperature. That is, it contains all the moisture it can hold at that temperature.

Total Heat. When the sensible heat and the latent heat are added together, the sum is known as the total heat. In air conditioning, 0°F (−17.8°C) is usually taken as the point from which heat content is measured. Total heat is also indicated by a wet bulb thermometer.

Saturated Gas (Vapor). When a gas or water vapor is at the temperature of the boiling point that corresponds to its pressure, it is also at its saturation temperature. This saturated state is the condition under which a gas exists above its liquid in a closed container, as in a refrigerant drum or an evaporator. A saturated gas contains no superheat.

Pound of Dry Air. When this term is used, it means 1 lb (0.4536 kg) of dry air. The following terms are examples of the use of the term *pound of air:* total heat per pound of air, moisture per pound of dry air, or latent heat per pound of air. In each case this means a pound of dry air and does not refer to a pound of the mixture.

Cubic Foot of Air. This term merely indicates 1 cu ft (0.028 m³) of air and is used in expressing the quantity of moisture present in 1 cu ft of dry air when it is saturated.

Ventilation. The process of supplying or removing air from a space is known as ventilation whether it is conditioned air, used air, or fresh air.

Effective Temperature. Effective temperature is a temperature that is determined by experiment. It is not a temperature measured on a thermometer, but is a measure of personal comfort as felt by an individual. Effective temperature is produced by the correct combination of dry bulb temperature, relative humidity, and air motion.

Comfort Zone. This is the range of temperature and humidity in which most people feel comfortable. The outer limits of the comfort zone are not clearly defined due to dependence on outside conditions.

Overall Coefficient of Heat Transmission. The quantity of Btu (cal) which is transmitted each hour through 1 sq ft (0.0929 m²) of any material, or combination of materials, for each degree of temperature difference between the two sides of a partition is known as the overall coefficient of heat transmission. Note that the difference between the two sides of the material is the difference in temperature of the air on both sides, not the difference in temperature between both surfaces of the material.

Static Pressure. Static pressure is the pressure that a gas or air exerts at right angles against the walls of an enclosure duct. Static head or static pressure of air is usually measured in inches (millimeters) of water column by use of manometers and pilot tubes. The frictional loss in air ducts is known as pressure drop or loss of static head.

Velocity Pressure. This is the pressure due to the air movement. Velocity pressure is created by the energy of motion or kinetic energy in the moving air. It is measured in inches (millimeters) of water column.

Total Pressure. Total pressure is the sum of the static pressure and velocity pressure and is a measure of the total energy of the air.

Stack Effect. This is an upward flow or draft created by the tendency of heated air to rise in a vertical or inclined duct. Warm air rises because it expands and becomes lighter when heated. This principle is of special importance in the heating of tall buildings, where the entire building acts as a chimney (stack) due to the warmer air inside it. The warm air rises and escapes out of openings or cracks on the upper floors, while cold air is drawn in through doors, windows, cracks, and so on, on the lower floors to replace the warm, rising air. As a result, extra radiation must be used on the lower floors to offset the chilling effect of the incoming cold air.

Plenum Chamber. The plenum chamber is an equalizing chamber or air supply compartment to which the various distributing ducts of an air conditioning or ventilating system are connected. The plenum chamber is maintained under pressure, which causes the air to flow into the different distribution ducts.

PSYCHROMETRICS

The science which deals with the relationships that exist within a mixture of air and water vapor is known as psychrometrics. In air conditioning, psychrometrics involves the measuring and determining of the different properties of air both inside and outside the conditioned space. Psychrometrics can also be used to establish the conditions of air that will provide the most comfortable conditions in a given air conditioning application.

PSYCHROMETRIC CHART

A psychrometric chart is a diagram which represents the various relationships that exist between the heat and moisture content of air and water vapor. Different air conditioning manufacturers have slightly different versions of the psychrometric chart, locating the various properties in different places on the chart. The factors shown on a complete psychrometric chart are dry bulb temperature, wet bulb temperature, dew point temperature, relative humidity, total heat, vapor pressure, and the actual moisture content of the air.

■ Identification of Lines and Scales on a Psychrometric Chart

The illustrations that follow will help to locate the different lines and scales on a psychrometric chart. Consider the psychrometric chart as being a shoe with the toe pointing to the left and the heel on the right (see Figure 1-6).

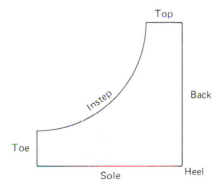

FIGURE 1-6
Psychrometric chart outline.

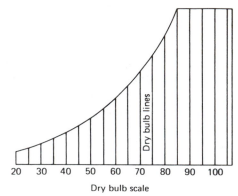

FIGURE 1-7
Dry bulb lines. (Courtesy of Research Products Corp.)

Dry Bulb Temperature Lines. The dry bulb temperature scale is located on the "sole" of the psychrometric chart (see Figure 1-7). The dry bulb lines extend vertically upward from the sole. There is one line for each degree of temperature.

Wet Bulb Temperature Lines. The wet bulb temperature scale is found along the "instep" of the chart, extending from the toe to the top (see Figure 1-8). The wet bulb lines extend from the instep diagonally downward to the right. There is one line for each degree of temperature.

Relative Humidity Lines. On a complete psychrometric chart, the relative humidity lines are the only curved lines on it (see Figure 1-9).

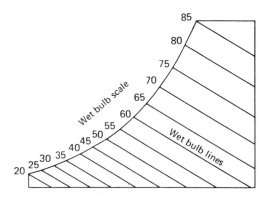

FIGURE 1-8
Wet bulb lines. (Courtesy of Research Products Corp.)

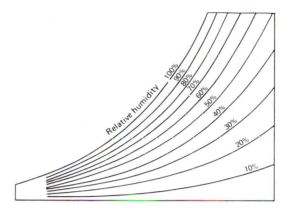

FIGURE 1-9
Relative humidity lines. (Courtesy of Research Products Corp.)

The various relative humidities are indicated on the lines themselves. There is no coordinate scale as with the other air properties.

Absolute Humidity Lines.   The scale for absolute humidity is a vertical scale on the right side of the chart (see Figure 1-10). The absolute humidity lines run horizontally to the left from this scale.

Dew Point Temperature Lines.   The scale for the dew point temperature is identical to the scale for the wet bulb temperature (see Figure 1-11). The dew point temperature lines run horizontally to the right—but not diagonally as in the case of wet bulb temperature lines.

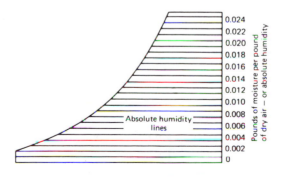

FIGURE 1-10
Absolute humidity lines. (Courtesy of Research Products Corp.)

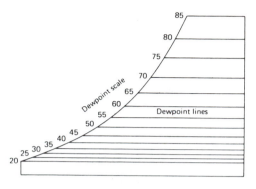

FIGURE 1-11
Dew point lines. (Courtesy of Research Products Corp.)

Specific Volume Lines.   The specific volume scale is located along the sole of the chart from 12.5 to 14.5 cu ft (0.3538 to 0.4103 m³) (see Figure 1-12). The cubic feet lines extend diagonally upward to the left from the sole to the instep of the chart. The specific volume lines on the chart are identified in terms of cubic feet per pound of dry air.

Enthalpy Lines.   The enthalpy scale is located along the instep of the chart (see Figure 1-13). The enthalpy lines are the same as the wet bulb lines on the psychrometric chart. Enthalpy is the total heat content. It can be used with the psychrometric chart to measure any heat change that takes place in any given psychrometric process. Both sensible and latent heat can be measured by use of enthalpy. It is a quick means of finding either of these factors.

When we put together the seven charts we have just covered, we have a complete psychrometric chart (see Figure 1-14).

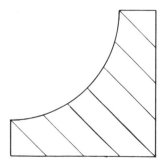

FIGURE 1-12
Cubic feet lines.

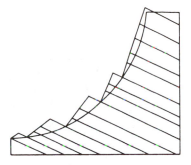

**FIGURE 1-13
Enthalpy lines.**

■ Use of the Psychrometric Chart

Any two of the foregoing properties which are located on the psychrometric chart will determine the condition of the air. If we choose any dry bulb temperature and any wet bulb temperature, the point at which these two lines intersect on the chart is the point that indicates the condition of the air at those given temperatures. The condition of the air at this point is definitely fixed. Similarly, the condition of the air at any other point on the psychrometric chart is fixed by the particular dry bulb and wet bulb temperatures. Because the possible combinations of any two temperatures are unlimited, there are an infinite number of possible conditions of air and an equally infinite number of possible points that may be plotted on the chart.

When a fixed air condition has been located at a point on the chart, all the other properties of that sample of air can be determined from the chart. Similarly, by use of the psychrometric chart, any two properties of the air and water vapor combination are sufficient to determine a condition of the air and all its other properties.

EXAMPLE 1:   Given a dry bulb temperature of 95°F (35°C) and a dew point temperature of 54°F (12.2°C), find the wet bulb temperature (see Figure 1-15).

Solution:

　　　Step 1.   Locate 95°F (35°C) on the dry bulb scale.

　　　　　　2.   Draw a line straight upward to the instep.

　　　　　　3.   Follow the instep down until 54°F (12.2°C) is found.

　　　　　　4.   Extend this point horizontally to the right until the 95°F (35°C) dry bulb line is intersected.

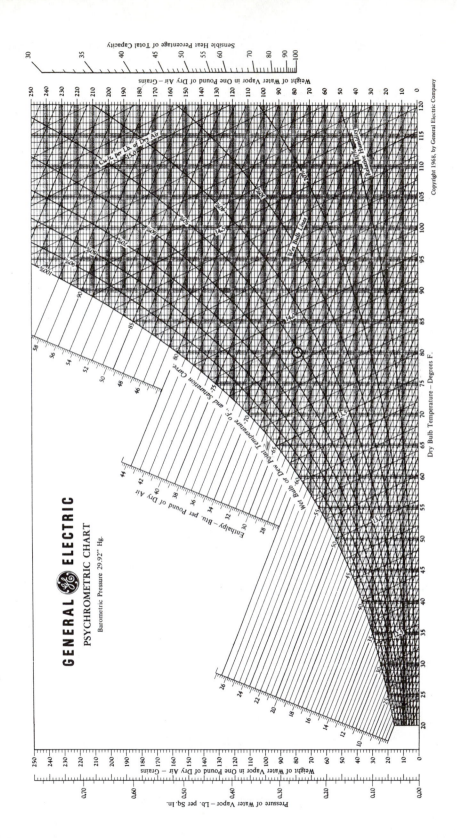

FIGURE 1-14
Psychrometric chart. (Courtesy of General Electric Central Air Conditioning Division)

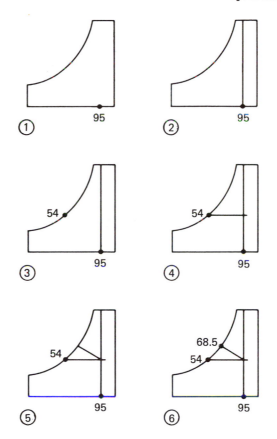

FIGURE 1-15
Finding wet bulb temperature.

5. Extend the point upward to the left to the wet bulb scale on the instep.

6. Read the wet bulb temperature of 68.5°F (20.3°C).

EXAMPLE 2:   Given a wet bulb temperature of 74°F (23.3°C) and a dew point temperature of 70°F (21.1°C), find the dry bulb temperature (see Figure 1-16).

Solution:

Step 1.   Locate 74°F (23.3°C) on the wet bulb scale.

2. Draw a line diagonally down the right to the back of the chart.

3. Locate 70°F (21.1°C) on the dew point scale.

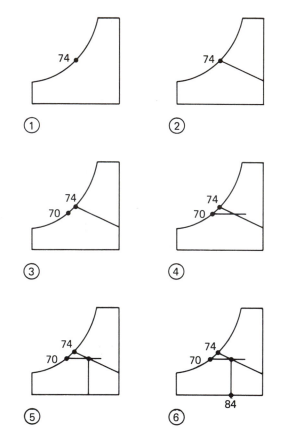

FIGURE 1-16
Finding dry bulb temperature.

4. Extend this point horizontally to the right until the wet bulb line is crossed.

5. Extend this point vertically downward until the dry bulb line is crossed.

6. Read the dry bulb temperature of 84°F (28.9°C).

EXAMPLE 3:   Given a wet bulb temperature of 73°F (22.8°C) and a dry bulb temperature of 81°F (27.2°C), find the dew point temperature (see Figure 1-17).

Solution:

Step 1.   Locate 81°F (27.2°C) on the dry bulb scale.

2.   Draw a line vertically upward to the instep of the chart.

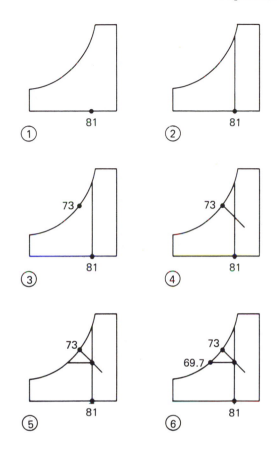

FIGURE 1-17
Finding dew point temperature.

3. Locate 73°F (22.8°C) on the wet bulb scale.
4. Extend this point diagonally downward until the dry bulb line is crossed.
5. Extend this point horizontally to the left to the instep.
6. Read a dew point temperature of 69.7°F (20.9°C).

EXAMPLE 4: Given a dry bulb temperature of 95°F (35°C) and a dew point temperature of 54°F (12.2°C), find the relative humidity (see Figure 1-18).

Solution:

Step 1. Locate 95°F (35°C) on the dry bulb scale.
2. Draw a line vertically upward to the instep of the chart.

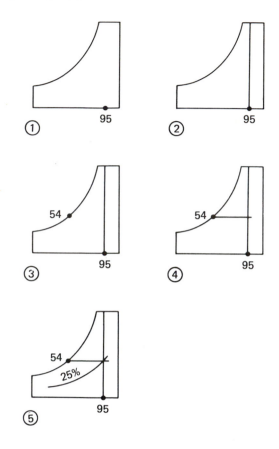

FIGURE 1-18
Finding relative humidity.

     3. Locate 54°F (12.2°C) on the dew point scale.

     4. Extend this point horizontally to the right until the dry bulb line is crossed.

     5. Read the relative humidity of 25% at this point.

EXAMPLE 5:   Given a wet bulb temperature of 74°F (23.3°C) and a dew point temperature of 70°F (21.1°C), find the total heat content of the air (see Figure 1-19).

Solution:

    Step 1. Locate 74°F (23.3°C) on the wet bulb scale.

         2. Draw a line diagonally upward to the left to the total heat scale.

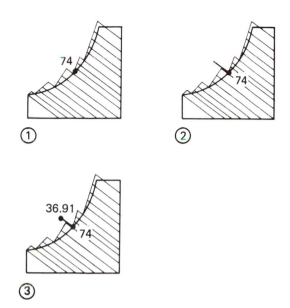

FIGURE 1-19
Finding total heat content using wet bulb
temperature.

3. Read the total heat to be 36.91 Btu (9.31 kcal) per pound (0.4536 kg) of air.

EXAMPLE 6: Given a dry bulb temperature of 95°F (35°C) and a dew point temperature of 54°F (12.2°C), find the total heat content of the air (see Figure 1-20).

Solution:

Step 1. Locate 95°F (35°C) on the dry bulb scale.

2. Draw a line vertically upward to the instep of the chart.

3. Locate 54°F (12.2°C) on the dew point scale.

4. Extend this point horizontally to the right until the dry bulb line is crossed.

5. Extend this point diagonally upward to the left to the total heat scale.

6. Find the total heat to be 32.32 Btu (8.14 kcal) per pound (0.4536 kg) of air.

This total heat is found by interpolation.

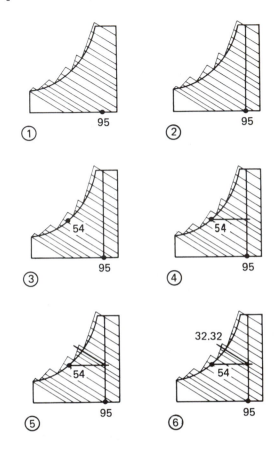

FIGURE 1-20
Finding total heat content using dry bulb temperature.

Total heat at 69°F (20.6°C) wet bulb = 32.71 Btu (8.24 kcal)
Total heat at 68°F (20°C) wet bulb = 31.92 Btu (8.04 kcal)
Difference = 0.79 Btu (0.199 kcal)
68.5° − 68° = 0.5°
0.5° × 0.79 = 0.395

Therefore, total heat at 68.5°F (20.3°C) wet bulb = 31.92 + 0.395 = 32.315 or 32.32 Btu (8.14 kcal).

■ Change in Total Heat

During the cooling season, we are interested primarily in the amount of heat that must be removed so that the air is cooled enough to fulfill the inside design conditions. In the heating season, heat is added to

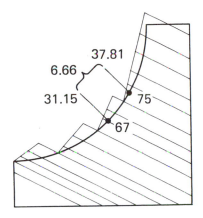

FIGURE 1-21
Change in total heat.

the air to meet the inside design conditions. Suppose, for instance, that the outside design condition is 75°F (23.9°C) wet bulb and it is desired to maintain a 67°F (19.4°C) wet bulb temperature inside the conditioned space. The amount of total heat that must be removed per pound (453.6 g) of dry air is found by the following method:

> Total heat at 75°F (23.9°C) wet bulb = 37.81 Btu/lb
> or 20.79 kcal/kg of air
> Total heat at 67°F (19.4°C) wet bulb = 31.15 Btu/lb
> or 17.13 kcal/kg of air
> Difference = 6.66 Btu/lb
> or 3.66 kcal/kg of air

Therefore, the total heat that must be removed in cooling air from 75°F (23.9°C) wet bulb down to 67°F (19.4°C) wet bulb is 6.66 Btu (1.68 kcal) per pound (453.6 g) of dry air (see Figure 1-21).

■ Sensible Heat

Sensible heat is heat that can be added to or removed from a substance without a change of state. If that substance is air, changing the sensible heat will result only in a change in temperature. The sensible heat content is indicated by the dry bulb temperature. A change in dry bulb temperature will result in a change in sensible heat only because there is no change of state.

EXAMPLE 7:   During the heating season air is to be heated from a 65°F (18.3°C) dry bulb temperature and a 50°F (10°C) wet bulb temperature to an 88°F (31.1°C) dry bulb temperature and a 60°F

(15.6°C) wet bulb temperature. Find the sensible heat that must be added per pound of dry air (see Figure 1-22).

Solution:

Step 1.   Locate 65°F (18.3°C) dry bulb and 50°F (10°C) wet bulb.

   2.   Locate 88°F (31.1°C) dry bulb and 60°F (15.6°C) wet bulb.

   3.   Extend these points diagonally upward to the left to the total heat scale.

   4.   Read 20.19 and 26.18 Btu (11.1 kcal/kg and 14.4 kcal/kg).

   5.   Obtain the difference:

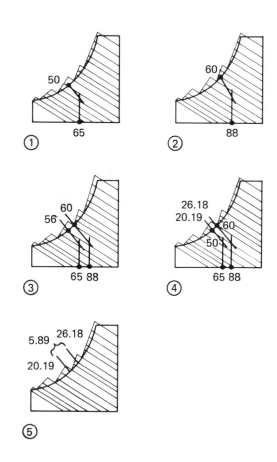

FIGURE 1-22
Sensible heat change on heating air.

Total heat at 60°F (15.6°C) wet bulb = 26.18 Btu/lb
(14.4 kcal/kg)
Total heat at 50°F (10°C) wet bulb = 20.19 Btu/lb
(11.1 kcal/kg)
Sensible heat added = 5.99 Btu/lb
(3.3 kcal/kg) of dry air

This heat change is sensible heat only because there was no change in the moisture content of the air.

■ Latent Heat

The dew point temperature is always an indication of the moisture content of the air. Any change in the dew point temperature will result in a change in moisture content. The moisture content can only be changed by changing the dew point temperature. Therefore, it should be noted that as long as the dew point temperature remains constant there will be no change in the moisture content or water vapor that is present in the air.

The moisture content scale used on most charts indicates grains of moisture per pound (453.6 g) of dry air. The dew point temperature is read on the same scale as the wet bulb temperature.

The latent heat of vaporization is the amount of heat in Btu (cal) required to change a liquid to a gas at a constant temperature. If we now have a given number of grains of moisture per pound (453.6 g) of dry air, there must have been a certain amount of heat required to vaporize this moisture into the air. This is the latent content of the air and water vapor mixture.

EXAMPLE 8:   Air at 75°F (23.9°C) dry bulb and 57°F (13.9°C) wet bulb is to be conditioned to obtain a 75°F (23.9°C) dry bulb temperature and a 70°F (21.1°C) wet bulb temperature. Find the amount of latent heat added and the grains per pound (453.6 g) of moisture added (see Figure 1-23).

Solution:

Step 1. Locate 75°F (23.9°C) dry bulb and 57°F (13.9°C) wet bulb.

2. Extend this point to the total heat scale and read 24.5 Btu (6.2 kcal).

3. Locate 75°F (23.9°C) dry bulb and 70°F (21.1°C) wet bulb.

4. Extend this point to the total heat scale and read 34 Btu (8.61 kcal).

5. The heat added to this air is 34 − 24.5 = 9.5 Btu (2.4 kcal) per pound (453.6 g) of dry air.

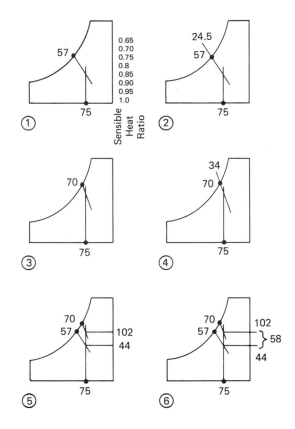

FIGURE 1-23
Latent heat change and grains of moisture added.

6. Extend these points to the right to the grains of moisture scale and read 44 grains (2.85 g) at the beginning of the process and 102 grains (6.61 g) of moisture per pound (453.6 g) of dry air at the end of the process.

7. The grains of moisture added per pound (453.6 g) of dry air is $102 - 44 = 58$ grains of moisture added.

This heat change is latent heat only because there was no change in the dry bulb temperature of the air.

■ Sensible Heat Ratio

Sensible heat ratio is the sensible heat removed in Btu (cal) divided by the total heat removed in Btu (cal). The sensible heat ratio indicates the sensible heat percentage of total heat removed. The sensible heat ratio

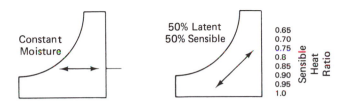

FIGURE 1-24
Sensible heat ratio line.

is above 50% in most comfort air conditioning. This is because most comfort air conditioning systems remove more sensible heat than latent heat.

The ratio will vary with the type of installation. A residence may have a 0.7 or 0.8 sensible heat factor, whereas a restaurant may have a 0.5 or 0.6 sensible heat ratio. This indicates that 70 or 80% of the total change in heat removed from the residential air is sensible heat.

The sensible heat ratio scale is located on the right-hand side of the psychrometric chart. When the conditions of air are plotted on the chart and a line is drawn through them and extended to the sensible heat ratio scale, the ratio is read directly (see Figure 1-24). When the line is at approximately a 45° angle, the sensible heat ratio is 50%, or 0.50. This indicates that half of the heat removed is sensible and half is latent.

EXAMPLE 9:   The desired room conditions are 80°F (26.7°F) dry bulb and 67°F (19.4°C) wet bulb. The supply air conditions are 60°F (15.6°C) dry bulb and 58°F (14.4°C) wet bulb. Find the sensible heat factor (see Figure 1-25).

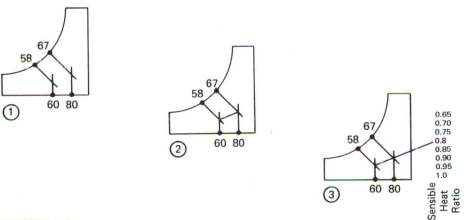

FIGURE 1-25
Finding sensible heat ratio (line method).

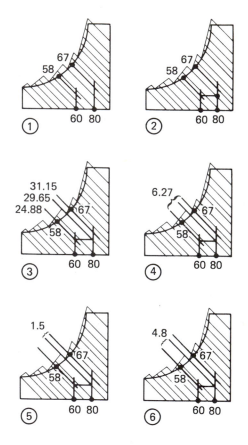

FIGURE 1-26
Calculated sensible heat ratio.

Solution:

Step 1. Plot the given conditions on the psychrometric chart.

2. Draw a straight line between the two points.

3. Extend the line to the sensible heat ratio scale.

4. Read 0.8 on the sensible heat ratio scale.

Another method of finding the sensible heat ratio is to use a ratio of the total heat removed to the sensible heat removed from the air. If we use the same conditions as those stated in Example 9, we can figure the sensible heat ratio as follows (see Figure 1-26).

Solution:

Step 1. Plot the given conditions on the psychrometric chart.

2. Extend the final air condition [60°F (15.6°C) dry bulb and 58°F (14.4°C) wet bulb] point horizontally to the right until the 80°F (26.7°C) dry bulb line is crossed.

3. Extend the 80°F (26.7°C) dry bulb line vertically downward to the sole of the chart. This forms a reverse "L."

4. Extend the room air point, the supply air point, and the intersection of the two sides of the "L" to the total heat scale.

5. Read the total heat for each line extended; room conditions, 31.15 Btu (7.85 kcal); supply conditions, 24.88 Btu (6.3 kcal); and intersection of "L," 29.65 Btu (7.5 kcal).

6. Calculate total heat removed: 31.15-24.88 = 6.27 Btu (1.5 kcal).

7. Calculate latent heat removed: 31.15-29.65 = 1.5 Btu (0.38 kcal).

8. Calculate sensible heat removed: 29.65-24.88 = 4.8 Btu (1.2 kcal).

9. Calculate sensible heat ratio: $\dfrac{4.8}{6.27} = 0.76$.

This indicates that 76% of the heat removed is sensible heat.

■ Mixing Two Quantities of Air at Different Conditions

The science of air conditioning involves the process of mixing air, such as when the conditioned air supply is mixed with the room air or when it is mixed with bypass air. Another example is when fresh air is supplied in some proportion for ventilation purposes. When the initial condition and the quantities of these two air sources are known, the condition of the final mixture can easily be found by use of the psychrometric chart.

EXAMPLE 10: Outdoor air at 95°F (35°C) dry bulb and 75°F (23.9°C) wet bulb, point A, is to be mixed with return air at 70°F (21.1°C) dry bulb and 10% relative humidity, point B. The mixture is to consist of 25% outdoor air and 75% return air. Find the resulting dry bulb and wet bulb temperatures of the mixture (see Figure 1-27).

Solution;

Step 1. Plot points A and B on the chart.

2. Draw a line between the points.

3. Locate the dry bulb temperature by adding the percent-

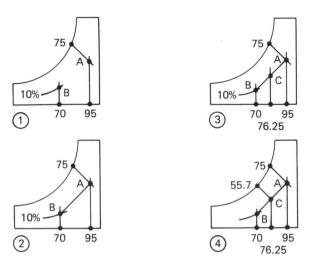

FIGURE 1-27
Mixing two quantities of air.

ages of each dry bulb temperature, that is, 25% of 95°F (13.3°C) = 23.75, 75% of 70°F (29.4°C) = 52.5. Resulting dry bulb temperature of mixture = 76.25°F (24.58°C).

4. Locate 76.25°F (24.58°C) dry bulb on the scale and extend this point to the mixture line, point C.

5. Extend point C to the wet bulb scale and read 55.7°F (13°C).

It is very important to note that it is permissible to find the resulting dry bulb temperature of a mixture of air by the percentage method, but it is not permissible to use this method for determining the resulting wet bulb temperature. For example, if the percentage method was used in the preceding example, the results would be as follows:

25% of 75°F = 18.75
75% of 47°F = 32.25
         51.00   incorrect resulting wet bulb temperature

The percentage method can be used to find the resulting wet bulb temperature by indirect application. To do this it is necessary first to find the total heat for each condition of the air and then apply the percentage method to find the resulting total heat of the air mixture. From the total heat of the mixture it is then possible to locate the corresponding correct wet bulb temperature.

EXAMPLE 11:   The total heat at 75°F (23.9°C) is 37.81 Btu (9.53 kcal). The total heat at 47°F (8.3°C) is 18.60 Btu (4.7 kcal). Find the corresponding wet bulb temperature.

Solution:

25% of 37.81 = 9.45 Btu (2.38 kcal)

75% of 18.60 = 13.95 Btu (3.52 kcal)

Total heat of moisture = 23.40 Btu (5.89 kcal)

The corresponding wet bulb temperature is 55.6°F (13.1°C).

■ Apparatus Dew Point and Air Quantity

On the psychrometric chart, the point where the sensible heat ratio line and the saturation curve intersect is known as the *apparatus dew point* (ADP). This point represents the lowest temperature at which air can be supplied to the conditioned space and still pick up the required amount of sensible and latent heat. If air at a higher ADP was supplied to the conditioned space, the quantity of air could be adjusted so that the required amount of sensible heat would be picked up. However, this air quantity would not pick up a sufficient amount of latent heat. The resultant air would then permit a higher relative humidity inside the conditioned space than would be desired. If the ADP was too low, the relative humidity also would be too low. However, if air at any condition that falls on the sensible heat ratio line between condition (point A) and ADP (point B) is supplied, the quantity of air can be adjusted so that the exact amount of both sensible and latent heat and the design conditions can be met (see Figure 1-28).

The ADP will actually approach very closely the cooling coil surface temperature. Therefore, a very large amount of coil surface would be needed if the conditioned air were to be supplied at the ADP. The air entering the cooling coil will be at its maximum temperature and the maximum air temperature differential will exist between the air and the coil surface. As the air progresses through the coil, it is cooled so that the temperature differential progressively decreases. The first row of tubing will do the most work and each following row will do less work. Therefore, to cool the air to the ADP would take an infinite number of tubes and the last rows would do very little work because the temperature differential that causes the heat transfer would be very small.

In actuality, the choice of the number of rows of tubing in depth usually falls between a minimum of three rows and a maximum of eight rows. When a large amount of sensible heat is to be removed, a smaller number of rows can be used. Conversely, as the latent heat

A — Room Air
B — Apparatus Dew Point
C — Supply Air
D — Outside Air
E — Mixture of A and D

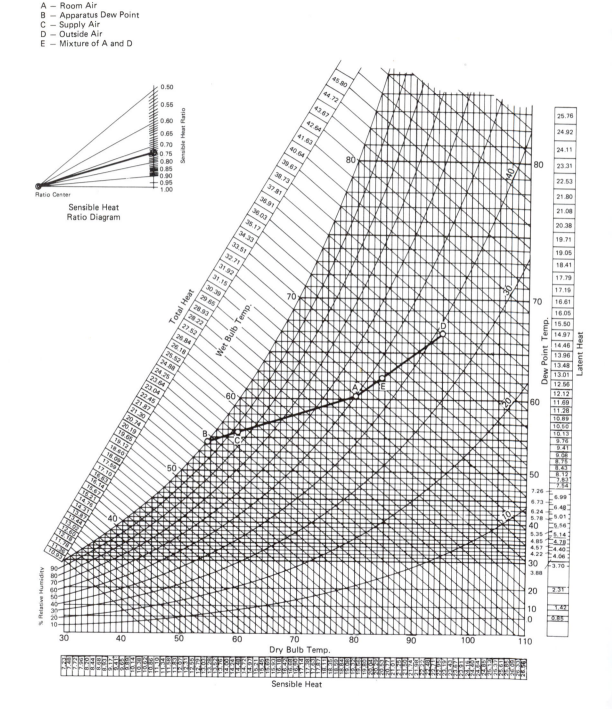

FIGURE 1-28
Psychrometric chart showing ADP.

32

load increases, more rows of tubing become necessary from both the engineering and the economic standpoints.

SUMMARY

The term *air conditioning* is defined as: the simultaneous mechanical control of temperature, humidity, air purity, and air motion. Unless all of these conditions are controlled, the term *air conditioning* cannot be properly applied to any system.

The two primary reasons for using air conditioning are (1) to improve the control of an industrial process, and (2) to maintain human comfort.

In the air conditioning process, there are three ways that the body gives off heat: (1) convection, (2) radiation, and (3) evaporation. In most cases the body uses all three methods at the same time.

There appears to be no set rule as to the best conditions for all people. In the same atmospheric conditions, a young person may be slightly warm whereas an elderly person may be cool.

The three conditions that affect the ability of the body to give off heat are (1) temperature, (2) relative humidity, and (3) air motion. A change in any one of these conditions will either speed up or slow down the cooling process.

Air at a temperature lower than that of the skin will speed up the convection process.

Relative humidity regulates the amount of heat the body can reject through radiation.

An increase in the rate of evaporation of perspiration from the body is largely a result of air movement. Evaporation is dependent on the ability of air to absorb moisture.

Dry air is air that contains no water vapor.

Absolute humidity is the actual quantity of weight of moisture present in the air in the form of water vapor.

Relative humidity is the ratio between the moisture content of a given quantity of air at any given temperature and pressure compared to the maximum amount of moisture which that same quantity of air is capable of holding at that temperature and pressure.

The process of adding moisture to the air is called humidification.

The process of removing moisture from the air is called dehumidification.

Saturated air is air that contains all the water vapor it is capable of holding at that particular pressure and temperature.

The temperature of the air as indicated by an ordinary thermometer is known as the dry bulb temperature and is a measure of the sensible heat content of the air.

The wet bulb temperature is the temperature of the air as measured by an ordinary thermometer with the bulb covered with a wet cloth or gauze.

The difference between the dry bulb temperature and the wet bulb temperature for any given condition is called the wet bulb depression.

The dew point temperature is the temperature at which water vapor will start to condense out of the air.

When air is at the dew point temperature, it is also at the saturation temperature and is considered to be saturated with moisture.

When the sensible heat and latent heat are added together, the sum is known as the total heat.

When a gas or water vapor is at the temperature of the boiling point which corresponds to its pressure, it is also at its saturation temperature.

When the term *pound of dry air* is used, it means 1 lb (0.4536 kg) of dry air and does not refer to a point of the air–moisture mixture.

The term *cubic foot of air* indicates 1 cu ft (0.028m³) of air and is used in expressing the quantity of moisture present in 1 cu ft of dry air when it is saturated.

The process of supplying or removing air from a space is known as ventilation whether it is conditioned air, used air, or fresh air.

Effective temperature is not a temperature measured on a thermometer, but is a measure of personal comfort as felt by an individual.

The comfort zone is the range of temperature and humidity in which most people feel comfortable.

The quantity of Btu (cal) which is transmitted each hour through 1 sq ft (0.0929 m²) of any material, or combination of materials, for each degree of temperature difference between the two sides of a partition is known as the overall coefficient of heat transmission.

Static pressure is the pressure that a gas or air exerts at right angles against the walls of an enclosure duct.

Velocity pressure is the pressue due to the air movement.

Total pressure is the sum of the static pressure and velocity pressure and is a measure of the total energy of the air.

Stack effect is an upward flow or draft created by the tendency of heated air to rise in a vertical or inclined duct.

The plenum chamber is an equalizing chamber or air supply compartment to which the various distributing ducts of an air conditioning or ventilating system are connected.

The science which deals with the relationships that exist within a mixture of air and water vapor is known as psychrometrics.

Sensible heat is heat that can be added to or removed from a substance without a change of state.

The dew point temperature is always an indication of the moisture content of the air. Any change in the dew point temperature will result in a change in moisture content.

Sensible heat ratio is the sensible heat removed in Btu (cal) divided by the total heat removed in Btu (cal). The sensible heat ratio indicates the sensible heat percentage of total heat removed.

On the psychrometric chart, the point where the sensible heat ratio line and the saturation curve intersect is known as the apparatus dew point (ADP). This point represents the lowest temperature at which air can be supplied to the conditioned space and still pick up the required amount of sensible and latent heat.

REVIEW QUESTIONS

1. Define total air conditioning.
2. Name the two primary reasons for using air conditioning.
3. Upon what is the comfort of a human body dependent?
4. Why is body heat rejection a constant process?
5. Name the three processes the body uses to give off heat.
6. What conditions relating to air are best for all people?
7. Name the three conditions that affect the ability of the body to give off heat.
8. What will happen to a person's body temperature if the air temperature is higher than the skin?
9. How will a high relative humidity affect body temperature?
10. Name three ways in which evaporation from the body can be increased.
11. Define dry air.
12. How is relative humidity determined?
13. What is the name of the process of adding moisture to the air?
14. What is air called that contains all the water vapor it is capable of holding at that particular pressure and temperature?
15. What device is used to measure the wet bulb temperature?
16. How is total heat determined?
17. What happens when the air temperature is lowered below the dew point temperature?
18. Define ventilation.
19. What is the term for the range of temperature and humidity in which most people feel comfortable?
20. What is the term for the sum of static pressure and velocity pressure?

21. What is the name of the science which deals with the relationships that exist within a mixture of air and water vapor?

22. What is a psychrometric chart?

23. Where is the dry bulb temperature scale located on the psychrometric chart?

24. What is the name for the only curved lines on the psychrometric chart?

25. Define sensible heat.

2 Heat Load Calculation Factors

The process of estimating the size of air conditioning and heating equipment required to serve a given space or building requires a working knowledge of the different sources of heat gain or loss through a structure. There are a great number of heat gains and losses through the structure and each must be considered when making these calculations.

INTRODUCTION

The primary function of air conditioning and heating equipment is to maintain indoor conditions that (1) aid in human comfort, (2) are required by a product, or (3) are required by a process performed in the space. To accomplish these conditions. equipment that has been properly sized must be installed and controlled through all seasons. The size of the equipment is calculated by the actual peak load requirements. As a general rule, it is all but impossible to measure accurately either the actual peak load or a partial load in any given space. Therefore, these loads are estimated: hence the term *heat load estimate.*

RESIDENTIAL HEATING AND COOLING LOAD ESTIMATE WORKSHEET

	ORIEN-TATION	TD COOL/HEAT	U FACTOR	ENTIRE HOUSE			LIVING ROOM			DINING ROOM			KITCHEN		
					Btu/hr			Btu/hr			Btu/hr			Btu/hr	
				Area	Cool	Heat	Area	Cool	Heat	Area	Cool	Heat	Area	Cool	Heat
WINDOWS															
1	N/S														
2															
3	NE/NW														
4															
5	E/W														
6															
7	SE/SW														
8															
WALLS AND PARTITIONS															
9															
10															
11															
12															
SUN LOAD															
13															
14															
15															
16															
FLOOR AND CEILING															
17															
18															
DOORS															
19															
20															
21															
22															
23															
24															
APPROXIMATE AIR QUANTITIES															
25															

FIGURE 2-1
Survey and checklist.

BATH 1			BEDROOM 1			BEDROOM 2			BEDROOM 3			DEN		
	Btu/hr			Btu/hr			Btu/hr			Btu/hr			Btu/hr	
Area	Cool	Heat	Area	Cool	Heat	Area	Cool	Heat	Area	Cool	Heat	Area	Cool	Heat

SURVEY

The first step before a heat load can be estimated is to make a comprehensive survey of the structure and immediate surroundings to assure an accurate evaluation of the heat load factors. When the building facilities together with the actual instantaneous heat load through a given part of the building are given careful consideration, accurate equipment selection can be made which will result in smooth, trouble-free performance. A survey and checklist is usually used to assure a thorough and accurate survey (see Figure 2-1).

HEAT SOURCES

Heat sources may be listed under two general headings: (1) those that result in an internal load on the conditioned space, and (2) those that result in an external load. The external heat load is a load on the indoor coil but does not affect the conditioned air after it is delivered to the space. The following outline classifies these sources:

1. Heat sources that result in an internal load:
   a. Conduction through walls, glass, roof, and so on
   b. Excess sun heat
   c. Duct heat gain
   d. Occupants
   e. Lighting load
   f. Equipment and appliances
   g. Infiltration of outside air
2. Heat sources that result in an external load:
   a. Ventilation air
   b. Any heat added to the air after it leaves a room

Heat that enters any given space may come from any or all of the sources listed above. A careful survey must be made on each job so that the heat load may be figured accurately. The function of an air conditioning system is to remove this heat gain, or add to the heat loss, and keep the inside at the desired condition. Care must be taken to ensure that the total load finally used is the load for the peak hour. Many of the heat gains, or losses, are not at a peak at the same time. For instance, in a restaurant the sun load may be at a peak at noon, whereas the peak customer load may not occur until around seven o'clock in the evening. It would be unnecessary to add these two peaks together. Again, it is possible that the lighting load may be a larger load than the sun load and not occur at the same time. From the foregoing it may be seen that the load on an air conditioning plant is subject to great variation. Therefore, the load is calculated on the basis of 1 hour, the hour of peak load, not on the basis of 24 hours. The cooling and heating system must be able to handle this peak hour load. Remember that we are looking for the 1 hour of the day when the sum of the loads is at a peak, not a sum of the peak loads.

CONDUCTION LOAD

Heat flows from a warm object to a cooler object. Therefore, if the temperature within a conditioned space is at a temperature lower than the outdoor temperature, or if the temperature within a conditioned space is at a temperature higher than the outdoor temperature, heat will be conducted through the walls, windows, and so on, to the cooler air. The amount of heat that will flow through a wall having different temperatures on the two sides depends on three factors:

1. The area of the wall

2. The heat-conducting characteristics of the wall, called the overall coefficient of heat transmission, commonly designated as U

3. The difference in temperature between the spaces separated by the wall

These same three factors also apply to windows, doors, roofs, and all building construction. The heat flow through a wall may be calculated by use of the following formula, which is basic for all problems involving heat flow:

$$Q = A \times U \times (T_1 - T_2)$$

where Q = heat transmission per hour, Btu

A = area of the wall, square feet

U = number of Btu that will flow through 1 sq ft of the given wall, in 1 hour, with a difference in temperature of 1°F between the two sides

T_1 = outside temperature

T_2 = inside temperature (T_1 and T_2 should be reversed when calculating heat gains)

The temperature differential, $T_1 - T_2$, is often given as TD, and when so written the formula becomes

$$Q = A \times U \times \mathrm{TD}$$

Table 2-1 gives a few of the U factors for various types of construction. A more complete list of these factors may be found in the *ASHRAE Guide and Data Book* published yearly by the American Society of Heating, Refrigerating and Air-Conditioning Engineers. The *Guide* presents the most accurate information to be had.

TABLE 2-1
Partial List of Heat Transmission Factors

Structure	U
EXTERIOR WALLS	
Wood siding, 2-in. × 4-in. studs, plastered	0.25
Wood siding, 2-in. × 4-in. studs, plastered, ½-in. insulation	0.17
Brick veneer, 2-in. × 4-in. studs, plastered	0.27
Brick veneer, 2-in. × 4-in. studs, plastered, ½-in. insulation	0.18
Stucco, 2-in. × 4-in. studs, plastered	0.30
Brick wall, 8-in. thick, no inside finish	0.50
Brick wall, 8-in. thick, furred, plastered on wood lath	0.30
Brick wall, 12-in. thick, furred and plastered on wood lath	0.24
Hollow tile, 10-in. thick	0.39
Hollow tile, 10-in. thick, furred, plastered on wood lath	0.26
4-in. brick hollow tile, 12-in. thick, furred, plastered on wood lath	0.20
4-in. brick-faced concrete, 10-in. thick	0.48
INTERIOR WALLS OR PARTITIONS	
Studding, wood lath and plaster both sides	0.34
Studding, wood lath and plaster both sides, ½-in. insulation	0.21
4-in. hollow tile, plastered both sides	0.40
4-in. common brick, plastered both sides	0.43
FLOORS, CEILINGS, AND ROOFS	
Plastered ceiling without flooring above	0.62
Plastered ceiling with 1-in. flooring	0.28
Plastered ceiling on 4-in. concrete	0.59
Suspended plastered ceiling under 4-in. concrete	0.37
Concrete 4-in. (floor or ceiling)	0.65
Concrete 8-in. (floor or ceiling)	0.53
Average wood floor	0.35
Average concrete roof	0.25
Average wood roof	0.30
GLASS AND DOORS	
Window glass, single	1.13
Double-glass windows	0.45
Doors, thin panel	1.13
Doors, heavy panel (1¼-in.)	0.59
Doors, heavy panel (1¼-in.) with glass storm door	0.38

When calculating the conduction heat load, each type of construction must be separated and figured by itself.

EXAMPLE 1:   Let us consider a wall 80 ft long and 12 ft high constructed of 8-in. brick, furred and plastered on wood lath. In this wall there is one thin panel door, 6 ft × 2½ ft. The temperature difference (TD) across the wall is 15°F. Calculate the conduction heat load.

Solution: From Table 2-1 we find that the U for this type of construction is 0.30 and for the thin panel door is 1.13.

Gross area of the wall: $80 \times 12 = 960$ sq ft
Area of the door: $6 \times 2½ = 15$ sq ft
Net area of the wall: 945 sq ft

Substituting the values in the formula $Q = A \times U \times TD$, we have

Wall: $945 \times 0.30 \times 15 = 4252.5$ Btu/hr
Door: $15 \times 1.13 \times 15 = \underline{254.25}$ Btu/hr
Total heat gain: 4506.75 Btu/hr

EXAMPLE 2:   Let us take a space that is somewhat more complicated. Study the sketch in Figure 2-2, then follow through the calculations.

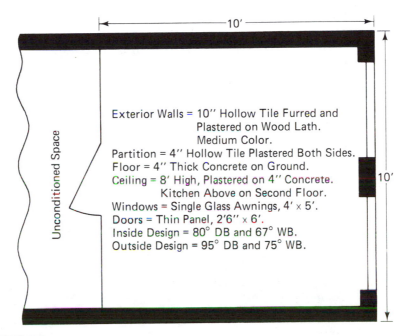

Exterior Walls = 10″ Hollow Tile Furred and
                Plastered on Wood Lath.
                Medium Color.
Partition = 4″ Hollow Tile Plastered Both Sides.
Floor = 4″ Thick Concrete on Ground.
Ceiling = 8′ High, Plastered on 4″ Concrete.
                Kitchen Above on Second Floor.
Windows = Single Glass Awnings, 4′ × 5′.
Doors = Thin Panel, 2′6″ × 6′.
Inside Design = 80° DB and 67° WB.
Outside Design = 95° DB and 75° WB.

FIGURE 2-2
Sketch for Example 2.

Solution: For the present we consider only the conduction heat gain through the exterior walls and glass.

Gross area of the exterior wall: 3 (10 ft × 8 ft) = 240 sq ft
Area of the exterior glass: 2(4 ft × 5 ft) = 40 sq ft
Net exterior of the wall: 200 sq ft

From Table 2-1 we find *U* for the wall: 0.26
U for single glass: 1.13
TD: 95°F − 80°F = 15°F

Then: Wall: 200 × 0.26 × 15 = 780 Btu/hr
Glass: 40 × 1.13 × 15 = 678 Btu/hr

This is all sensible heat because it was calculated on a difference in dry bulb temperatures.

TEMPERATURE DIFFERENCES

In computing the conduction heat load, the 15°F differential for exterior walls in practice does not remain constant for all surfaces during the cooling season. For example, if the conditioned room is located on

TABLE 2-2
Temperature Differential for Abnormal
Conditions (Based on 15°F Normal Differential
for Exterior Walls)

STRUCTURE	D
Walls—exterior	15
Glass in exterior walls	15
Glass in partitions	10
Partitions next to unconditioned rooms	10
Partitions next to laundries, kitchens, boiler rooms, or store show windows having a large lighting load	30
Floors above unconditioned rooms	10
Floors on ground or over unvented spaces or basements	0
Floors above laundries, kitchens, or boiler rooms	40
Floors above vented spaces	15
Ceilings with unconditioned rooms above	10
Ceilings with laundries, kitchens, etc., above	25
Ceiling with roof directly above (no attic)	15
Ceiling with totally enclosed attic and roof above	20
Ceiling with cross-vented attic and roof above	15

NOTE: If normal differential is 20°F, add 5°F to values above.

the opposite side of a partition from a boiler room or laundry which has high temperatures, it is obvious that a 15°F differential is not sufficient. In such cases, where a 15°F differential for the exterior walls is normal for the installation, the differential across the partition would be much higher than 15°F. Table 2-2 gives the temperature difference in degrees that should be used if the normal differential for the exterior walls is 15°F. *If the normal is to be maintained at 20°F, differential for the exterior walls then add 5°F to the temperature differences given in the table.* For example, ceilings with laundries or kitchens above show an indicated differential of 25°F when the normal is 15°F. If we use a normal differential of 20°F, then we must add 5°F to the 25°F, which gives a differential of 30°F for this particular ceiling.

EXAMPLE 2 *(continued):*   As an example of the use of Table 2-2, let us return to Figure 2-2 and calculate the heat gain through the partition, floor, and ceiling.

Solution:

Gross area of the partition:	10 ft × 8 ft =	80 sq ft
Area of door:	6 ft × 2½ ft =	15 sq ft
Net area of partition:		65 sq ft

From Table 2-1 we find

U for partition:	0.40
U for thin panel door:	1.13

From Table 2-2 we find that partitions next to unconditioned rooms have a temperature differential of 10°F, when the normal is 15°F, which we have already found is the case in this problem. This same differential also applies to the door. We now have

Partition: 65 × 0.40 × 10 = 260 Btu/hr
Door: 15 × 1.13 × 10 = 169.5 Btu/hr

Area of the floor:	10 ft × 10 ft =	100 sq ft
Area of the ceiling:	10 ft × 10 ft =	100 sq ft

From Table 2-1:

U for the floor: 0.65
U for the ceiling: 0.59

From Table 2-2:

TD for the floor: 0°F
TD for the ceiling: 25°F

Then:

Ceiling: 100 × 0.59 × 25 = 1475 Btu/hr
Floor: 100 × 0.65 × 0 =   0 Btu/hr

We can now find the total conduction heat leakage by adding together all the values found.

Exterior walls	=	780.0
Exterior glass	=	678.0
Partition	=	260.0
Door in partition	=	169.5
Ceiling	=	1475.0
Floor	=	0
Total load		3362.5 Btu/hr

This means that each hour the air conditioning unit must remove 3362.5 Btu of heat gained by conduction because this is the amount of heat that will be conducted through the walls and other surfaces. The other sources of heat, which must also be removed by the air conditioning unit, will be taken up as we go along.

DESIGN TEMPERATURES

The outdoor temperature in various parts of the country undergoes considerable day-to-day variation. For example, a northern city such as Minneapolis, Minnesota, may have extremely high temperatures during periods of the summer season, but on the average the temperatures in Minneapolis are lower than those in Dallas, Texas. In determining the conduction heat load for a cooling or heating unit, it is necessary to choose a design temperature on which to base the calculations. Weather Bureau records show that extreme temperatures occur in most localities during less than 10% of the time in the cooling and heating seasons. Therefore, the design temperature for comfort air conditioning is not based on the highest or lowest temperature occurring in that locality. It is based on the average maximum temperature. Obviously, it would not be economical to install the much larger plant that would be required to furnish full cooling or heating capacity to take care of temperatures that occur only a small percentage of the time. Table 2-3 is for rough calculations and it is not recommended that this table be used when accurate figures are desired.

TABLE 2-3
Design Outside Dry Bulb Temperatures

LOCATION IN UNITED STATES	TEMPERATURE (°F)
Northern portion	90
Central portion	95
Southern portion	100

Design dry bulb temperatures for various cities throughout the United States are given in Table 2-4. For all our work these are the temperatures to be used in estimating. Table 2-4 also shows the design wet bulb for cooling and design dry bulb for winter. For example, notice that the correct outside summer design conditions for Detroit, Michigan, would be 95°F dry bulb and 75°F wet bulb, with a winter design condition of −10°F.

The design conditions when established as discussed above will prove satisfactory for all comfort air conditioning work even though there may be times during any year when they may be exceeded. At such times there will be a rise in either or both the dry bulb temperature or relative humidity within the conditioned space. This rise is usually small and in addition is normally of short duration, so that larger equipment is generally not economically justified in comfort conditioning. This, however, is not true of industrial air conditioning systems that are installed for the benefit of the product or process. In this case, the loss of design conditions within the conditioned space for even a few hours might result in the loss of thousands of dollars worth of products or even completely shut down some processes. It can be seen that a few such losses could far exceed the cost of the larger air conditioning equipment needed to prevent them. It is therefore economically sound to design industrial systems far closer to the maximum outside conditions that will be experienced. As a general rule, industrial systems should be designed for a 5°F greater outside dry bulb and a 3°F greater outside wet bulb than the standard design conditions for comfort air conditioning.

SUN EFFECT OR EXCESS SUN LOAD

When a roof, ceiling, wall, or window is exposed to the direct rays of the sun, the surface will warm up rapidly, giving the effect of a greater differential between the outside and inside surfaces. There are several factors that affect the quantity of heat which will be gained due to the sun effect:

1. Time of day
2. The direction in which the exposure faces
3. The color of the exposed surface (such as color of wall or roof)
4. The type of surface (that is, rough or smooth)
5. The latitude

The time of day is very important because the sun effect varies greatly from one hour to the next. By means of the chart, Figure 2-3,

TABLE 2-4
Design Temperature Conditions (°F) Used for
Calculating Heat Loads for Heating or Cooling
for Various Regions of the United States

| State | City | Extreme Temperatures | | Mean Temperatures | | Design conditions | | |
| | | | | | | Winter | Summer | |
		Low	High	January	July	Dry bulb	Dry bulb	Wet bulb
Ala.	Mobile	− 1	103	52	81	15	95	79
Ariz.	Phoenix	16	119	51	90	25	110	75
Ark.	Little Rock	−12	108	41	81	5	96	78
Calif.	San Francisco	27	101	50	58	30	90	65
Colo.	Denver	−29	105	30	72	−10	95	72
Conn.	New Haven	−14	101	28	72	0	95	75
D. C.	Washington	−15	106	33	77	0	96	78
Fla.	Jacksonville	10	104	55	82	30	95	79
Ga.	Atlanta	− 8	103	43	78	5	95	78
Idaho	Boise	−28	121	30	73	−10	100	70
Ill.	Chicago	−23	103	25	74	−10	95	75
Ind.	Indianapolis	−25	106	28	76	− 5	96	76
Iowa	Dubuque	−32	106	19	74	−15	96	75
Kan.	Wichita	−22	107	31	79	0	100	75
Ky.	Louisville	−20	107	34	79	0	98	77
La.	New Orleans	7	102	54	82	25	96	80
Maine	Portland	−21	103	22	68	−10	92	75
Md.	Baltimore	− 7	105	34	77	0	96	78
Mass.	Boston	−18	104	28	72	− 5	95	75
Mich.	Detroit	−24	104	24	72	−10	95	75
Minn.	St. Paul	−41	104	12	72	−20	95	75
Miss.	Vicksburg	− 1	104	48	81	15	96	80
Mo.	St. Louis	−22	108	31	79	0	98	79
Mont.	Helena	−42	103	20	66	−20	90	68

the excess differential, due to the sun, for any wall or roof may be found for any hour of the day. Note that the sun effect on the east wall is greatest at 10:00 A.M., on the south walls and the roof at 2:00 P.M., and on the west walls at 6:00 P.M. When the sun's rays are perpendicular, the rate of heat absorption is greatest, but due to the time lag in passing through the wall, it is not effective in the conditioned space until sometime later. This time lag will vary depending on the construction of the surface. For instance, a light frame construction may have a lag of less than an hour, whereas a heavy masonry construction may have a lag of several hours. For average construction, a time

TABLE 2-4 (Continued)

State	City	Extreme Temperatures		Mean Temperatures		Design conditions		
						Winter	Summer	
		Low	High	January	July	Dry bulb	Dry bulb	Wet bulb
Neb.	Omaha	−32	111	22	77	−15	100	75
Nev.	Winnemucca	−28	104	29	71	−10	95	70
N. C.	Charlotte	− 5	103	41	78	10	96	79
N. D.	Bismarck	−45	108	8	70	−25	98	70
N. H.	Concord	−35	102	22	68	−15	95	75
N. J.	Atlantic City	− 7	104	32	72	0	95	75
N. M.	Santa Fe	−13	97	29	69	0	92	70
N. Y.	New York City	−14	102	31	74	0	95	77
Ohio	Cincinnati	−17	105	30	75	0	98	77
Okla.	Oklahoma City	−17	108	36	81	0	100	76
Ore.	Portland	− 2	104	39	67	10	95	70
Penna.	Philadelphia	− 6	106	33	76	0	95	78
R. I.	Providence	−12	101	29	72	0	95	75
S. C.	Charleston	7	104	50	81	15	96	80
S. D.	Pierre	−40	112	16	75	−20	100	72
Tenn.	Nashville	−13	106	39	79	5	98	79
Texas	Galveston	8	101	54	83	25	95	78
Utah	Salt Lake City	−20	105	29	76	− 5	95	70
Vt.	Burlington	−28	100	19	70	−15	92	73
Va.	Norfolk	2	105	41	79	10	98	78
Wash.	Seattle	3	98	40	63	10	90	67
W. Va.	Parkersburg	−27	106	32	75	− 5	96	77
Wis.	Milwaukee	−25	102	21	70	−15	94	75
Wyo.	Cheyenne	−38	100	26	67	−15	92	70

(Courtesy of American Society of Heating, Refrigerating and Air-Conditioning Engineers, Inc., Atlanta, Ga.)

lag of 2 hours is usually taken. Thus, at noon the sun's rays are perpendicular to the roof, but the peak effect is not felt until 2:00 P.M.

There is no time lag through window glass because the sunlight passes instantly through the glass and gives up heat when it strikes objects in the room. Sunlight heat gain is maximum during the time when the sun shines through the window with greatest intensity. When a building is to be air conditioned, it is common practice to shade the window with awnings, which will reduce the effect of the sun to a considerable extent.

Dark surfaces on such areas as walls and roofs absorb considerably more heat than do light-colored surfaces. For this reason, roofs are sometimes given a coat of aluminum paint to reduce the quantity of heat transferred due to sun radiation. Sun effect can almost be omitted when there is a ventilated air space, attic, or room above the space being cooled, or when the roof is sheltered or shaded by adjacent buildings or trees except for a very short period at noon. If partly shadowed, figure the exposed part only. Spraying the roof with water will cool the surface and under such conditions the sun effect may be omitted. Sun effect on skylights and window glass may be omitted if the surface is sheltered from the sun's rays by adjacent walls or trees. If a portion of the surface is exposed, figure the portion that is exposed.

The peak sun load in one-story buildings usually occurs at a time when the sun effect on the roof is at a peak. However, under conditions where steps have been taken to reduce the sun load on the roof, such as painting the roof with aluminum paint, this statement does not necessarily hold true. On multiple-story buildings where there is a room above, or where steps have been taken to control sun effect, as in one-story buildings, the peak sun load may occur at some hour in the forenoon or afternoon. Ordinarily, the areas of the east and west walls and glass, and to some extent the south walls and glass, are the governing factors. Great care must be exercised in determining the sun load. Good common sense used in inspection and determination of the peak load is an absolute necessity.

In calculating the excess sun load, do not forget that it is figured separately from the conduction load and is an excess over the conduction load.

The sun load is calculated exactly like the conduction load; that is, the area is determined and multiplied by the same U factor, which is a property of the wall. This is then multiplied by the solar temperature differential found from the chart.

EXAMPLE 2 *(continued)*:   Let us now find the excess sun load on the east wall and glass of Figure 2-2.

Solution:

Gross area of wall: 10 ft × 8 ft = 80 sq ft
Area of glass: 2(4 ft × 5 ft) = 40 sq ft
New wall area: 40 sq ft

As before, the U factor for the wall is 0.26 and for the glass 1.13. From Figure 2-3 we find that the peak sun load on a medium-colored east wall would occur at 10:00 A.M., and from Figure 2-4 we find that the peak sun load on east glass, with awnings, will be at 8:00 A.M. We must determine when the combined sun load will be at a peak. This is done as in Figure 2-5. Notice particularly that the combination may

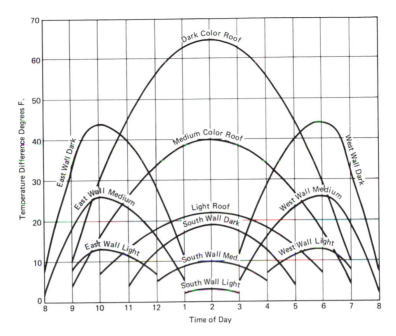

FIGURE 2-3
Solar chart.

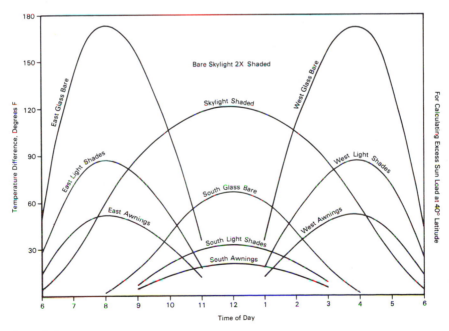

FIGURE 2-4
Solar chart.

	8:00 A.M.	9:00 A.M.	10:00 A.M.
East wall 40 × 0.26	× 2° = 20.8	× 18° = 187.2	× 26° = 270.4
East glass 40 × 1.13	× 52° = 2350.4	× 46° = 2079.2	× 32° = 1446.4
Total	2371.2	2266.4	1716.8

FIGURE 2-5
Example, calculation.

give a peak which does not come at the same time as either one of the separate peaks. By this tabulation we have found that the peak sun load on this east wall and glass will be 2371.2 Btu/hr at 8:00 A.M. This heat load would be in addition to the regular conduction load, as it would occur only when the sun was shining, whereas the conduction load would be in effect even though the sun was not shining. As we have said before, extreme care must be exercised in finding the peak hour.

OCCUPANCY

The heat load for people at rest is different from the load when people are active (see Table 2-5). These are about the only classifications needed for practical comfort conditioning calculations. Notice that this is the first item that has been split up into sensible and latent heat. The use of this table simply involves determining how many people fall into each classification and multiplying this number by the proper figure from the table.

As should be noted, the quantity of heat given off by the human body is dependent on the activity of the individual. For example, people seated at rest, as in the home, in restaurants, or in theaters, give off approximately 400 Btu/hr total heat. In the case of medium activity, this may increase to 675 Btu/hr. This load is partly sensible and partly latent, as shown in Table 2-5. Note that the sensible heat remains nearly constant.

TABLE 2-5
Body Heat of Occupants (Btu)

CONDITION	SENSIBLE	LATENT	TOTAL
Seated or at rest	225	175	400
Medium activity	225	450	675

EXAMPLE 3:   In a restaurant having 5 employees and a capacity of 30 customers, compute the body heat load using the factors given in Table 2-5.

Solution:

Sensible Heat:	$35 \times 225 =$	7,875 Btu
Latent heat:	$30 \times 175 =$	5,250 Btu
	$5 \times 450 =$	2,250 Btu
Total body heat:		15,375 Btu

(7875 Btu sensible + 7500 Btu latent)

ELECTRIC LIGHTS AND APPLIANCES

Less than 5% of the electrical energy supplied to electric lights is transformed into light and the other 95% is given off in the form of heat. Therefore, electric lighting within the conditioned space should be figured to the full amount of the electrical energy or 3.41 Btu/W. Only the electric lights that are on during the peak period should be included in the figure. If lighting is supplied by skylights or windows which make it unnecessary to use electric lights during the peak load period, the electric light factor should be omitted. A restaurant located on the main floor of a building of several stores in the downtown section would ordinarily be subject to very little sun effect and it is likely that the lights would be on throughout the day. Under such conditions, the lighting heat load must be included in the calculation. Cafeterias located in basements must be treated in the same manner. Where the installation is of such a nature that the sun will shine into the windows and artificial lighting is required at the same time, both sun effect and heat dissipated by electric lighting must be included in the calculation.

The heat load from incandescent lights is 3.41 Btu/hr per watt. Fluorescent lights have a heat load of 3.41 Btu/W plus the ballast. Thus approximately 25% should be added to the heat load values for fluorescent lights.

Electric appliances such as toasters and hair dryers must be figured the same as electric lights, on the basis of 3.41 Btu/W. If these devices dissipate all of their heat output within the conditioned space, their total combined wattage may be multiplied by 3.41 to obtain the heat output in Btu per hour. The length of time and the period during which the appliance operates must also be considered. For example, a toaster in a restaurant would be more apt to operate steadily during the breakfast period, but would probably not be used continually during the noon hour and period of peak load. In any event, the exact

TABLE 2-6
Heat from Electric Appliances

ITEM	AMOUNT OF HEAT
Lights	3.41 Btu/W
Appliances (toasters, hair dryers, etc.)	3.41 Btu/W
Motors ⅛–½ hp ¾–3 hp 5–20 hp	 4250 Btu/hp per hour 3700 Btu/hp per hour 3000 Btu/hp per hour

length of time and period of operation of these appliances should be ascertained from the owner.

Electric motors give off varying quantities of heat. The larger the motor, the greater the efficiencies and the lower the percentage of heat generated and given off in relation to the output or work performed. Allowances must also be made for motors that are within the conditioned space and operate the air conditioning equipment. The heat equivalent values for several different sizes of motors and electric appliances operated in a conditioned space are listed in various tables (see Table 2-6). Note that the values for motors are given on the basis of Btu per horsepower per hour.

GAS AND STEAM APPLIANCES

Appliances that include food cooking and warming devices, such as steam tables and coffee urns, are also listed in tables giving their heat load in Btu per hour (see Table 2-7). Note that for each pound of steam condensed by the coil in the steam table, 960 Btu must be added to the heat load. Half of this load is given off as sensible heat and half as latent heat, due to the water vapor given off by the food. Coffee urns are generally heated by gas applied directly. Under such conditions, it is necessary to figure the heat generated by the burning of the gas. The products of combustion of a gas are chiefly carbon dioxide and water vapor. Thus the heat load should be figured as approximately half sensible and half latent. The quantity of heat given off by coffee urns can be accurately calculated by figuring 1 cu ft of natural gas or 2 cu ft of manufactured gas for each gallon of rated capacity.

Where exhaust hoods are used over the urns and steam tables, the sensible and latent loads from these sources may each be reduced by 50%.

TABLE 2-7
Heat from Gas and Steam Appliances

Item	Btu/hr		
	Sensible	Latent	Total
Steam tables, per sq ft top surface	1,000	1,000	2,000
Restaurant coffee urns	5,000	5,000	10,000
Natural gas, per cu ft	500	500	1,000
Manufactured gas, per cu ft	275	275	550
Steam condensed in warming coils, per pound	480	480	960

NOTES: 1. If there are exhaust hoods over steam tables and coffee urns, include only 50% of the heat load from these sources.
2. Coffee urns may also be figured on the basis of consuming approximately 1 cu ft of natural gas or approximately 2 cu ft of manufactured gas per hour per gallon rated capacity.

INFILTRATION

Natural infiltration will occur in nearly every room unless pressure is maintained in the room, since an airtight room is almost impossible to build. In a residence or small office where there are no excessive tobacco fumes or other objectionable odors, natural infiltration may supply sufficient fresh air if there are window and skylight openings in excess of, or at least equal to, 5% of the floor area. In addition, the room must be sufficiently large to allow 50 sq ft of space and at least 500 cu ft of volume for each occupant. When these conditions are fulfilled and the infiltration is sufficient for the fresh air requirements, forced ventilation will not be necessary unless required by local codes and ordinances (see Table 2-8). The better types of installations rarely depend on infiltration for the required fresh air.

The amount of infiltration through windows and doors has been tabulated for convenience in Table 2-9. If there are two exposed walls, use the wall having the greater amount of leakage. If there are more than two exposed walls, use either the wall having the greatest amount of leakage or half of the total of all the walls, whichever quantity is larger.

EXAMPLE 4: Let us determine the infiltration into a room having three exposed walls. One of the exposed walls has three average-fit weather-stripped double-hung windows 2 ft high. The opposite wall has one industrial-type steel sash window, 3 ft × 3 ft, and the third wall has a poorly fitted door which is weather-stripped, measuring 3

TABLE 2-8
Minimum Outdoor Air Required for Ventilation
(Subject to Local Code Regulations)

Application	CFM per person
Apartment or residence	10–15
Auditorium	5–7½
Barber shop	10–15
Bank or beauty parlor	7½–10
Broker's board room	25–40
Church	5–7½
Cocktail lounge	20–30
Department store	5–7½
Drugstore	7½–10
Funeral parlor	7½–10
General office space	10–15
Hospital rooms (private)	15–25
Hospital rooms (wards)	10–15
Hotel room	20–30
Night clubs and taverns	15–20
Private office	15–25
Restaurant	12–15
Retail shop	7½–10
Theater (smoking permitted)	10–15
Theater (smoking not permitted)	5–7½

ft × 6 ft. The wind velocity averages 5 miles per hour (mph) during the summer season.

Solution: To determine the total length of the crack around the double-hung windows, use the following procedure:

Cracks at the top, center, and bottom of the window:	2 ft × 3 = 6 ft
Vertical cracks on each side of the window:	5 ft × 2 = 10 ft
Total crack length for one double-hung window:	16 ft
Total crack length for three windows:	16 ft × 3 = 48 ft

With a wind velocity of 5 mph for an average-fit weather-stripped double-hung window produces an infiltration of 5 cu ft per hour per foot of crack (see Table 2-9).

Thus 48 ft × 5 = 240 cu ft of air per hour infiltration in the first wall.

The steel sash window has a total crack length as follows: 3 ft × 4 = 12 ft.

The infiltration for an industrial steel sash window at a 5-mph wind velocity is 50 cu ft/hr (see Table 2-9).

Thus 12 ft × 50 = 600 cu ft infiltration per hour in the second wall.

The infiltration through the door in the third wall is determined as follows:

$$\text{Total crack length around door} = (6 \text{ ft} \times 2) + (3 \text{ ft} \times 2) = 18 \text{ ft}$$

A poorly fitted weather-stripped door admits an infiltration of 30 cu ft/hr for a 5-mph wind velocity (see Table 2-9).

Thus 18 ft × 30 = 540 cu ft infiltration per hour in the third wall.

Since the total infiltration in the three walls equals 1380 cu ft, one-half of the infiltration would be 690 cu ft. This quantity, however, is greater than the infiltration through any individual wall. Therefore, this is the quantity used in determining whether or not there is sufficient fresh air being supplied to the room through infiltration. Since the air conditioning plant must treat all the air entering by infiltration, this must be added to the load on the plant. However, if the quan-

TABLE 2-9
Infiltration per Foot of Crack per Hour in Cubic Feet

Type of Opening	Condition	Wind Velocity (mph)					
		5	10	15	20	25	30
Double-hung wood window	Average fit, not weather-stripped	6	20	40	60	80	100
Double-hung wood window	Average fit, weather-stripped	5	15	25	35	50	65
Double-hung wood window	Poor fit, weather-stripped	6	20	35	50	70	90
Double-hung metal window	Not weather-stripped	20	45	70	100	135	170
Double-hung metal window	Weather-stripped	6	18	30	44	58	75
Steel sash	Casement, good fit	6	18	30	44	58	75
Steel sash	Casement, average fit	12	30	50	75	100	125
Steel sash	Industrial type	50	110	175	240	300	375
Doors	Good fit, not weather-stripped	30	70	110	155	200	250
Doors	Good fit, weather-stripped	15	35	55	75	100	125
Doors	Poor fit, not weather-stripped	55	140	225	310	400	500
Doors	Poor fit, weather-stripped	30	70	110	155	200	250

NOTE: Use of storm sash permits a 50% infiltration reduction for poorly fitting windows only.

tity of air that enters by infiltration is found to be insufficient for the minimum requirements of the room (see Table 2-8), then sufficient fresh air must be introduced into the air conditioning apparatus to take care of the requirements. This extra quantity of fresh air will prevent infiltration by supplying an extra amount of air into the conditioned space, building up internal pressure causing the air to leak outward rather than inward. This arrangement is obviously superior since all the air that enters the conditioned space has first been through the conditioning apparatus.

■ Infiltration Heat Load

It should be apparent that air which filters into a room being conditioned will be an internal heat load on the room. It must therefore be broken up into its sensible and latent components. This is not true of ventilation air, which is a load on the air conditioning apparatus but not on the room. When the air is supplied as ventilation, it is only necessary to figure the total heat load.

EXAMPLE 5: The amount of infiltration that we figured for Example 4 was 690 cu ft/hr. Since the basis of the psychrometric chart that we use to find our heat contents is 1 lb of dry air, we must change cubic feet per hour to pounds per hour. Find the sensible and latent components.

Solution: Let us assume that the outside air is at 95°F dry bulb and 75°F wet bulb, which are the design conditions for New York City (see Table 2-4). Let us assume that the inside design conditions are 80°F dry bulb and 76°F wet bulb. From the psychrometric chart, we find that dry air at 95°F dry bulb weighs 0.0715 lb/cu ft (see Figure 2-6). Thus

$$690 \times 0.0715 \text{ lb} = 49.335 \text{ lb of dry air per hour}$$
$$\text{Total heat at 75°F wet bulb} = 37.81 \text{ Btu/lb of dry air}$$
$$\text{Total heat at 67°F wet bulb} = \underline{31.15} \text{ Btu/lb of dry air}$$
$$\text{Total heat removed:} \quad 6.66 \text{ Btu/lb of dry air}$$

Thus $49.335 \times 6.66 = 328.57$ Btu/hour. This is the total heat to be removed due to the infiltration. It must now be separated into sensible and latent heat loads.

$$\text{Sensible heat content at 95°F dry bulb} = 22.95 \text{ Btu/lb of dry air}$$
$$\text{Sensible heat content at 80°F dry bulb} = \underline{19.32} \text{ Btu/lb of dry air}$$
$$\text{Sensible heat to be removed:} \quad 3.63 \text{ Btu/hr}$$

Thus $49.335 \times 3.63 = 179.09$ Btu/hr. This is the sensible heat to be removed. Total heat less sensible heat equals latent heat. Therefore, $328.57 - 179.09 = 149.48$ Btu/lb is the latent heat to be removed.

This calculation is similar to the examples dealing with the psychrometric chart in Chapter 1. We suggest that you review Chapter 1 at this time.

VENTILATION HEAT LOAD

The amount of natural infiltration should be compared with the minimum outside air requirements as determined from Table 2-8. If found to be insufficient, ventilation air must be introduced into the building. This introduction of ventilation air will prevent infiltration; therefore, we do not include both loads in the same job. In Example 4, we found that 690 cu ft of air per hour entered the room by infiltration. If the room is used as an office and has two occupants, we would find that 1200 cu ft of air per hour would be required since each occupant requires 10 cubic feet per minute (cfm) (see Table 2-8). Thus $2 \times 10 \times 60 = 1200$ cu ft/hr. Therefore, infiltration would not supply sufficient fresh air, and we would have to figure the heat load due to the 1200 cu ft/hr.

In this case we would not need to calculate the sensible and latent heat loads, but would just calculate the total load. This load in Btu per hour would be calculated exactly as the infiltration heat load. That is, we would first convert the cubic feet per hour to pounds of dry air per hour, and then multiply this by the difference in total heat content between the two samples of air as found from the psychrometric chart, Figure 2-6.

DUCT HEAT LOAD

In most cases, the heat gain or loss through the ducts is so small compared to the total load that it may be neglected. However, where ducts must be run through kitchens or other warm spaces or cooled spaces, it is necessary to figure the heat load. It should be apparent that any heat added to the air, from the time it leaves the evaporator coil until it gets back to the coil, is a heat load that must be removed by the evaporator coil. Similarly, any heat lost through the ducts must be added by the heating unit. The heat transfer factors are provided in tabular form for convenience (see Table 2-10).

It is almost impossible to figure the duct load in advance because the ducts are not sized when the rest of the heat load is figured and the unit selected. This usually necessitates redesign. It is obvious that in the case where the entire supply duct is located within the conditioned space, no duct heat load need be calculated since any heat

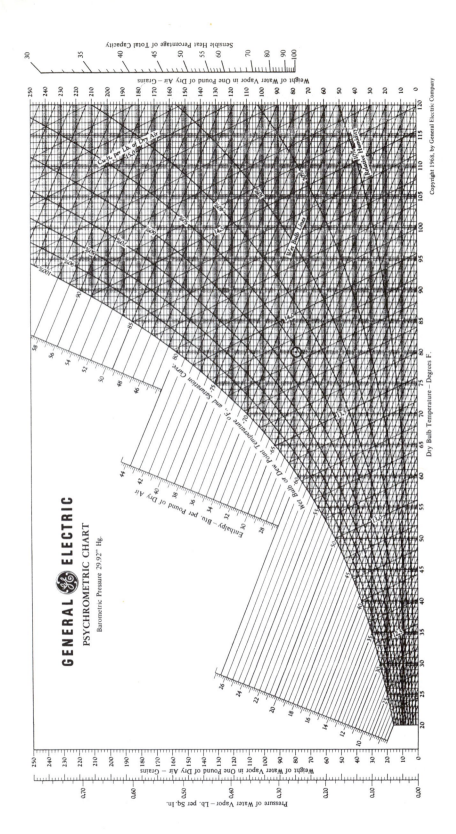

FIGURE 2-6
Psychrometric chart. (Courtesy of General Electric Air Conditioning Central Division)

General Electric
PSYCHROMETRIC CHART
Barometric Pressure 29.92" Hg.

Copyright 1968, by General Electric Company

TABLE 2-10
Heat Transfer Factors for Ducts

Duct	Btu/Sq Ft per Hour per 1°F Difference
Sheet metal, not insulated	1.13
Average insulation, ½ in. thick	0.41

pickup or loss must come from or be given up to the conditioned space. This then becomes a credit on the amount of cooling or heating required in the conditioned space in the exact amount of the heat gain or loss through the ducts.

Since the amount of air that must be supplied to the space is determined from the amount of sensible heat that must be removed, or added, it follows that the duct sizes are indirectly related to the sensible heat. It is therefore, very realistic to figure the duct heat load as a percentage of the sensible heat load. In actual practice, this percentage will vary from a minimum of zero for the case mentioned above to a maximum of about 5%. In making such an assumption, the length of the duct run through the unconditioned spaces, as well as the temperature of the air surrounding the duct, must be given consideration. Also, the percentages mentioned above are based on the assumption that all ducts running through unconditioned spaces will be provided with insulation.

When calculating the duct heat load using Table 2-10, the procedure is the same as for any other heat flow calculation. The area of exposed duct is found and multiplied by the factor selected from the table. This total is then multiplied by the difference between the dry bulb temperature of the air inside the duct and the dry bulb temperature of the air outside the duct.

FAN HEAT LOAD

In order to maintain the circulation of the conditioned air, it is necessary to furnish energy to the air. This is accomplished by the fan, which in turn derives its energy from its motor. Thus work is done on the air, with the result that its temperature will be raised in direct proportion to the amount of work done. In other words, the heat equivalent of the power necessary to drive the air is added to the air and this amount of heat must be removed by the refrigeration unit. This amount of heat is a plus during the heating season. It is heat that need not be supplied by the heating unit. Again, as in the case of the duct heat load, the fan power may not be known at the time the calculations are made. But like that allowance, the fan power is dependent on the amount of air circulated, which in turn is dependent on the total sensible heat. Here again the fan power allowance may be realistically taken as a percentage of the sensible heat. On the normal job this will amount to from 3 to 4% of the total sensible heat. In actual practice, it is customary to allow 5% as a combined factor for duct heat gain and fan power. This is adequate for the average job, but where special conditions prevail, suitable allowances must be made. On most equipment package arrangements, the fan heat load is considered in the Btu rating of the package.

SAFETY FACTOR

Because of the large number of variables involved, heat load calculation is not an exact science in the same sense as pure mathematics or chemistry. For example, heat transmission coefficients are determined with great accuracy for certain types of construction, yet the engineer can have no definite assurance that the construction of the space under consideration is identical in all respects to the test panels from which the coefficients were established. Also, such internal loads as people, lights, equipment, shading factors, and color of external surfaces may all vary from those on which the design is based. It is therefore reasonable and proper to apply a factor of safety to the calculated heat loads. This factor of safety should vary from a maximum on very small jobs where any variation may seriously affect the performance of the system to a minimum on very large jobs, where there is greater chance of the possible variations canceling one another. For the average job a factor of safety of 10% is normally sufficient. It must be noted that the same safety factor must be applied to both the sensible and latent heat loads in order not to upset the relationship between these two quantities. The importance of this will become more apparent after more experience has been gained in making these calculations. It must also be noted that the application of a safety factor in no way relieves the engineer from making as careful and thorough a survey and calculation as possible.

SUMMARY

The process of estimating the size of air conditioning and heating equipment required to serve a given space or building requires a working knowledge of the different sources of heat gain or loss through a structure.

The primary function of air conditioning and heating equipment is to maintain indoor conditions that (1) aid in human comfort, (2) are required by a product, or (3) are required by a process performed in the space.

The actual size of the equipment is calculated by the actual peak load requirements.

The first step before a heat load can be estimated is to make a comprehensive survey of the structure and immediate surroundings to assure an accurate evaluation of the heat load factors.

A survey and checklist is generally used to ensure a thorough and accurate survey.

Heat sources may be listed under two general headings: (1) those that result in an internal load on the conditioned space, and (2) those that result in an external load.

Many of the heat gains or losses are not at a peak at the same time.

Remember that when calculating the size of unit we are looking for the one hour of the day when the sum of the loads is at a peak, not a sum of the peak loads.

When calculating the conduction heat load, each type of construction must be separated and figured by itself.

In computing the conduction heat load, the 15°F differential for exterior walls in practice does not remain constant for all surfaces during the cooling season.

In determining the conduction heat load for a cooling or heating unit, it is necessary to choose a design temperature on which to base the calculations.

The design temperature for comfort air conditioning is not based on the highest or lowest temperature occurring in that locality. It is based on the average maximum temperature.

Industrial air conditioning, however, is usually designed to meet the maximum temperatures that normally occur in a given area.

When a roof, ceiling, wall, or window is exposed to the direct rays of the sun, the surface will warm up rapidly, giving the effect of a greater differential between the outside and inside surfaces.

The time of day is very important because the sun effect varies greatly from one hour to the next.

Because of the time lag in passing through a wall, the sun's rays are not effective in the conditioned space until sometime later.

There is no time lag through window glass because the sunlight passes instantly through the glass and gives up heat when it strikes objects in the room.

The peak sun load in one-story buildings usually occurs at a time when the sun effect on the roof is at a peak.

Ordinarily, the areas of the east and west walls and glass, and to some extent the south walls and glass, are the governing factors.

Great care must be exercised in determining the sun load. In calculating the excess sun load, do not forget that it is figured separately from the conduction load and is an excess over the conduction load.

The heat load for people at rest is different from the load when people are active.

Electric lighting within the conditioned space should be figured to the full amount of the electrical energy, or 3.41 Btu/W. Only the electric lights that are on during the peak period should be included in the figure.

When the installation is of such nature that the sun will shine into the windows and artificial lighting is required at the same time, both sun effect and heat dissipated by the electric lighting must be included in the calculation.

Electric appliances such as toasters and hair dryers must be figured the same as electric lights, on the basis of 3.41 Btu/W.

Electric motors give off varying quantities of heat. The larger the motors, the greater the efficiency and the lower the percentage of heat generated and given off in relation to the output or work performed.

Appliances that include food cooking and warming devices, such as steam tables and coffee urns, also give off heat to the conditioned space.

Natural infiltration will occur in nearly every room unless pressure is maintained in the room, because an airtight room is almost impossible to build.

Natural infiltration may supply sufficient fresh air if there are windows and skylight openings in excess of, or at least equal to, 5% of the floor area.

In most cases, the heat gain or loss through the ducts is so small compared to the total load that it may be neglected. However, where ducts must be run through hot or cold places, it is necessary to figure the heat load.

It is reasonable and proper to apply a safety factor to the calculated heat loads, and for the average job this should be about 10%.

REVIEW QUESTIONS

1. What is the primary function of air conditioning and heating equipment?
2. How is the size of equipment calculated?
3. What is the first thing to be done before estimating a heat load?
4. Under what two general headings are heat sources listed?
5. Are all of the heat gains or losses at a peak at the same time?
6. Is the heat load calculated on the basis of 1 hour or 24 hours?
7. What formula is used to calculate the heat flow through a wall?

8. Are comfort air conditioning and heating units sized for the maximum temperature encountered in that locality?

9. When estimating a heat load, is only the dry bulb temperature considered?

10. Is there any difference in designing an industrial unit and a residential unit?

11. What effect in differential between the inside and outside surface of a wall will the sun cause when shining on a building?

12. Is the time of day important when considering the sun effect on a building?

13. Will dark or light surfaces increase the heat load on a building?

14. Is there a time lag when the sun shines on window glass?

15. When considering occupant load, are people at rest considered differently from those who are active?

16. Upon what is the quantity of heat given off by a person dependent?

17. What percentage of the energy to an electric light is transformed into heat?

18. Does the amount of heat given off by an incandescent light equal the amount of heat given off by the same-size fluorescent light?

19. In regard to heat calculations, what should be done regarding air that filters into a conditioned space?

20. What is the maximum heat load that should be added for the duct work?

3

Heat Load Calculation Procedure

In this chapter we will calculate the heat load for a residential building. This calculation will cover all the heat gain and heat loss factors taken up in Chapter 2.

INTRODUCTION

As stated earlier, the primary function of air conditioning and heating equipment is to maintain indoor conditions that (1) aid in human comfort, (2) are required by a product, or (3) are required by a process performed in the space. To accomplish these conditions, equipment that has been properly sized must be installed and controlled through all seasons. The size of the equipment is calculated by the actual peak load requirements. In this chapter we will go through the process of sizing a heating and cooling unit for a residential building.

PROCEDURE FOR CALCULATING HEAT GAIN

In presenting this procedure we will calculate the heat gain for a single-story residential building. The type of construction is listed below under the heading "Construction Details for Figure 3-1."

While completing this heat gain calculation, refer to the text paragraphs, figures, and tables indicated. The information found there will aid in each step while progressing through the example. The assumed design conditions, orientation, type of construction, and so on, should be indicated on the survey and checklist, Figure 3-2. Refer to Figure 3-1 for the information required on the survey and checklist.

This example is presented only as an illustration for the use of data and the procedures used in calculating the building heat gain. Therefore, it is not to be considered as a recommendation for construction details.

■ Construction Details for Figure 3-1

- Walls: Brick veneer, 2 in. × 4 in. studs, plastered, ½-in. insulation
- Windows: Double-glass windows with light shades
  Window sizes: $W_1 = 3$ ft × 6 ft, $W_2 = 3$ ft × 5 ft,
  $W_3 = 3$ ft × 4 ft, $W_4 = 3$ ft × 3 ft

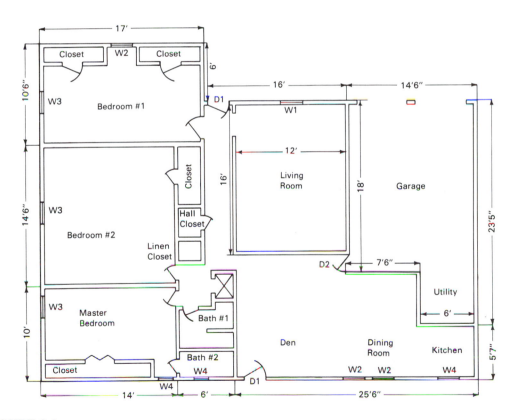

FIGURE 3-1
Model house.

SURVEY AND CHECKLIST

1. Design conditions:

   _____   _____
   (Summer)   *(Winter)*

   - Outside temp.: _____

   - Inside temp.: _____

   - Daily temp. range: _____

2. Building faces: N___ NE___ E___ SE___ S___ SW___ W___ NW___

3. Construction, type: Single-story___ Two-story___ Split-level___

   - Walls: Frame___ Heavy masonry (over 10 in. thick) ___

     Light masonry (under 10 in. thick) ___

   - Insulation: None___ 1 in.___ 2 in.___ 3⅝ in.___

   - Color: Light___ Dark___

   - Ceiling heights: 1st floor___ 2nd floor___

   - Roof: Pitched roof___ Flat roof___ White___ Dark___ Attic fan___

   - Ceiling: Studio___ Rafters covered___ Rafters exposed___

     Insulation: None___ 1 in.___ 2 in.___ 3⅝ in.___

   - Floor: Slab on ground___ Edge insulation: None___ 1 in.___ 2 in.___

     Over crawl space___Open___ Vented___ Closed: Unvented___ Vapor seal___

     Insulation: Underneath floor___ Thickness___ Crawl space walls___ Thickness___

     Over basement___ Over garage___

     Insulation: None___ 1 in.___ 2 in.___ 4 in.___

   - Windows: Movable___ Fixed___ Wood___ Metal___ Single-pane___ Double-pane___

     Glass block___ Double-hung___ Casement___ Weather-stripped___ Plain___

4. Type of shading: Location

N	NE	E	SE	S	SW	W	NW

FIGURE 3-2

Roof overhang (in.): _____

Awnings: _____

Outside shade screen: _____

Inside blinds or shade: _____

Garage or carport: _____

• Other types of shading: _____

5. Miscellaneous load items—
   Customer's specific requirements:

   • Frequent large-group entertainment: _____

   • Special temperatures required: Summer___ Winter___

   • Other: _____

6. Equipment location:

   • Air-cooled condenser: _____

   • Condensate pump required: _____

   • Indoor unit location: Basement___ Utility room___ Attic___

7. Utilities:

   • Electric service: Volts___ Phase___ Cycles___ Capacity of service___

   • Gas service: Natural___ Mixed___ Manufactured___ LP___ Heating value___ Size___

                   LP tank location___ Meter location___

   • Oil service: Tank location___ Above oil pump___ Below oil pump___ Distance___

                   Tank size___ Fill and vent location___ Fuel line location___

   • Condensate disposal: Sump___ Dry well___ Gravits___ Pump___ Floor drain___ Other_____

8. Existing installation data: Forced air___ Gravity___

   • Hot water___ Steam___ Other___

   • Make of furnace_____ Capacity_____

   • Blower wheel dimensions: Length___ Width___ Belt drive___ Direct drive___

                   Blower pulley size___ Blower motor pulley size___

(Continued on following page)

- Blower motor: Make_____ HP___ Full-load amps___ Phase___ Cycle___ Frame___

- Ductwork: Inadequate___ Adequate___

- Chimney: Type_____ Size___

- Comments: _____

- Controls: Standard___ Remote___ Automatic changeover___

    Continuous fan operation_____ Night setback_____

- Cost estimate—miscellaneous items:

    Electrical: Distance from unit: Inside___ Outside___

        Included in estimate:

            Yes___ No___

        Fused disconnect switches included in estimate:

            Yes___ No___

        Control wiring included in estimate:

            Yes___ No___

FIGURE 3-2   (Continued)

- Doors: Heavy panel with glass storm door
  Door sizes: $D_1 = 3$ ft 0 in. $\times$ 6 ft 8 in., $D_2 = 2.5$ ft $\times$ 6 ft 8 in.
- Roof: Light color
- Ceiling: Plastered without flooring above
- Floor: Concrete slab 4 in. thick
- Partition: Studding, wood lath and plaster both sides, ½-in. insulation
- Ducts: Located in attic with ½-in. insulation

The house is located in Dallas, Texas.

Complete the heat load chart by using the procedures and examples outlined below (see Figure 3-3). Fill in the required information under the headings "Orientation," "Construction Type," and "U Factor."

The first part of the construction on the heat load chart is the windows. Referring to "Construction Details for Figure 3-1," we find that the windows are of the double-glass type. In Table 3-1, at the bottom under "Glass and Doors," find that double-glass windows have a U factor of 0.45. Enter these items in the proper column on the chart. Since this table shows no difference in the direction the window is facing, we need to make only one entry (see Figure 3-3).

TABLE 3-1
Partial List of Heat Transmission Factors

Structure	U
EXTERIOR WALLS	
Wood siding, 2-in. × 4-in. studs, plastered	0.25
Wood siding, 2-in. × 4-in. studs, plastered, ½-in. insulation	0.17
Brick veneer, 2-in. × 4-in. studs, plastered	0.27
Brick veneer, 2-in. × 4-in. studs, plastered, ½-in. insulation	0.18
Stucco, 2-in. × 4-in. studs, plastered	0.30
Brick wall, 8-in. thick, no inside finish	0.50
Brick wall, 8-in. thick, furred, plastered on wood lath	0.30
Brick wall, 12-in. thick, furred and plastered on wood lath	0.24
Hollow tile, 10-in. thick	0.39
Hollow tile, 10-in. thick, furred, plastered on wood lath	0.26
4-in. brick hollow tile, 12-in. thick, furred, plastered on wood lath	0.20
4-in. brick-faced concrete, 10-in. thick	0.48
INTERIOR WALLS OR PARTITIONS	
Studding, wood lath and plaster both sides	0.34
Studding, wood lath and plaster both sides, ½-in. insulation	0.21
4-in. hollow tile, plastered both sides	0.40
4-in. common brick, plastered both sides	0.43
FLOORS, CEILINGS, AND ROOFS	
Plastered ceiling without flooring above	0.62
Plastered ceiling with 1-in. flooring	0.28
Plastered ceiling on 4-in. concrete	0.59
Suspended plastered ceiling under 4-in. concrete	0.37
Concrete 4-in. (floor or ceiling)	0.65
Concrete 8-in. (floor or ceiling)	0.53
Average wood floor	0.35
Average concrete roof	0.25
Average wood roof	0.30
GLASS AND DOORS	
Window glass, single	1.13
Double-glass windows	0.45
Doors, thin panel	1.13
Doors, heavy panel (1¼-in.)	0.59
Doors, heavy panel (1¼-in.) with glass storm door	0.38

Walls are the next item to be considered on the chart. The construction details list the walls as brick veneer, 2 ft × 4 ft studs, plastered, ½-in. insulation. Referring to Table 3-1 under the heading "Exterior Walls," we find the U factor to be 0.18. Enter these items on the chart.

RESIDENTIAL HEATING AND COOLING LOAD ESTIMATE WORKSHEET

	ORIENTATION	TD COOL/ HEAT	U FACTOR	ENTIRE HOUSE	Btu/hr		LIVING ROOM	Btu/hr		DINING ROOM	Btu/hr		KITCHEN	Btu/hr	
				Area	Cool	Heat	Area	Cool	Heat	Area	Cool	Heat	Area	Cool	Heat
WINDOWS															
1	N/S														
2		20/	0.45				18	162							
3	NE/NW														
4															
5	E/W														
6															
7	SE/SW														
8															
WALLS AND PARTITIONS															
9	Wall	20/	0.18				90.02	324.07					44.8	161.28	
10															
11	Part.	20/	0.21				127.35	534.87		104	436.8		48	201.6	
12															
SUN LOAD															
13	East wall	18/	0.18							30	97.2		39	126.36	
14	East glass	78/	0.45							30	1053		9	315.9	
15	East door	78/	0.38												
16															
FLOOR AND CEILING															
17	Ceiling	20/	0.62				248	3075.2		88	1091.2		33.6	416.64	
18	Floor	20/	0.65				248	3224		88	1144		33.6	436.8	
DOORS															
19	West	20/	0.38				19.98	151.85							
20															
21		Heat loss subtotal													
22	Internal cooling load	20% extra													
23	Cooling load	Heat loss			41045.58			7471.99			3822.7			1658.58	
24	%Total load														
APPROXIMATE AIR QUANTITIES															
25															

FIGURE 3-3
Survey and checklist.

BATH 1			BEDROOM 1			BEDROOM 2			BEDROOM 3			DEN		
Area	Btu/hr Cool	Btu/hr Heat	Area	Btu/hr Cool	Btu/hr Heat	Area	Btu/hr Cool	Btu/hr Heat	Area	Btu/hr Cool	Btu/hr Heat	Area	Btu/hr Cool	Btu/hr Heat
			27	243		12	108		12	108				
			241	867.6		104	374.4		68	244.8		16.65	69.93	
									142	460.08		76.02	2463	
									18	631.8				
												19.98	592.2	
60	744		178.5	2213.4		246.5	3056.6		158	1959.2		162	2008.8	
60	780		178.5	2320.5		246.5	3204.5		158	2024		162	2106	
	1524			5644.5			6743.5			5427.88			6647.73	

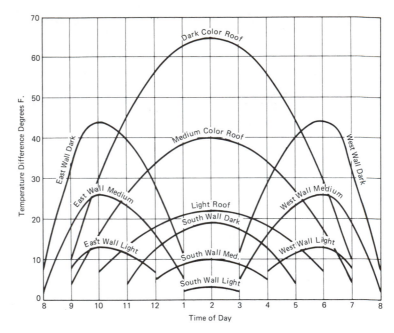

FIGURE 3-4
Solar Chart.

The sun load is somewhat more involved than the previous items. Checking Figure 3-4, we see that the roof heat gain would be at about 2:00 P.M. However, when we consider the brick walls as being a medium color value, the maximum load on the east walls would be about 6:00 P.M. and the south wall peak would be about 2:00 P.M. Since the peak load is greater at either 10:00 A.M. or 6;00 P.M., we can omit calculations for any other periods. Observing Figure 3-5 for the glass sun load, we find that the peak loads are at about 8:00 A.M. and 4:00 P.M. We can now determine the temperature difference for these periods and place them in a simple table to help in making the calculations.

	10:00 A.M.	6:00 P.M.
East wall	26° TD	26° TD
West wall		

	8:00 A.M.	4:00 P.M.
East glass	87° TD	
West glass		87° TD

Notice that there is no temperature differential difference between these two peak load times. Therefore, the wall with the most area and which has the largest area of glass will determine when the peak load occurs. Reviewing Figure 3-1, we can determine that the greatest load will be on the east side of the building. We can construct a chart showing the times between the peak wall load and the peak glass load to determine when the greatest load occurs and what that value is. We will use the formula $Q = A \times U \times TD$ to make these calculations.

	Area	U Factor	TD	8:00 A.M. Btu	TD	9:00 A.M. Btu	TD	10:00 A.M. Btu
West wall	287	×0.18×	2	103.32	18	929.99	26	1343.16
East glass	57	×0.45×	87	2231.55	78	2000.7	51	1308.15
East door	20	×0.38×	87	661.2	78	592.8	51	387.6
TOTALS				2996.07		3523.38		3038.91

We have determined that the peak sun load occurs on the east wall at 9:00 A.M. We can now enter these factors in the proper areas on the heat load chart (see Figure 3-3). Notice that we entered the TD on the chart because it will be different from the rest of the building. We will later divide this heat load among the various rooms on the east side of the building. Also, since doors are not shown in Figure 3-4 or 3-5 we assumed the same TD as that for the glass.

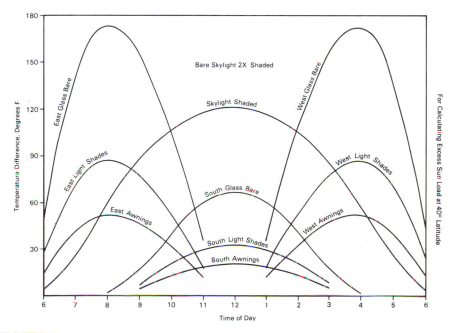

FIGURE 3-5
Solar chart.

TABLE 3-2
Design Temperature Conditions (°F) Used for
Calculating Heat Loads for Heating or Cooling
for Various Regions of the United States

State	City	Extreme Temperatures		Mean Temperatures		Design conditions		
						Winter	Summer	
		Low	High	January	July	Dry bulb	Dry bulb	Wet bulb
Ala.	Mobile	− 1	103	52	81	15	95	79
Ariz.	Phoenix	16	119	51	90	25	110	75
Ark.	Little Rock	− 12	108	41	81	5	96	78
Calif.	San Francisco	27	101	50	58	30	90	65
Colo.	Denver	− 29	105	30	72	− 10	95	72
Conn.	New Haven	− 14	101	28	72	0	95	75
D. C.	Washington	− 15	106	33	77	0	96	78
Fla.	Jacksonville	10	104	55	82	30	95	79
Ga.	Atlanta	− 8	103	43	78	5	95	78
Idaho	Boise	− 28	121	30	73	− 10	100	70
Ill.	Chicago	− 23	103	25	74	− 10	95	75
Ind.	Indianapolis	− 25	106	28	76	− 5	96	76
Iowa	Dubuque	− 32	106	19	74	− 15	96	75
Kan.	Wichita	− 22	107	31	79	0	100	75
Ky.	Louisville	− 20	107	34	79	0	98	77
La.	New Orleans	7	102	54	82	25	96	80
Maine	Portland	− 21	103	22	68	− 10	92	75
Md.	Baltimore	− 7	105	34	77	0	96	78
Mass.	Boston	− 18	104	28	72	− 5	95	75
Mich.	Detroit	− 24	104	24	72	− 10	95	75
Minn.	St. Paul	− 41	104	12	72	− 20	95	75
Miss.	Vicksburg	− 1	104	48	81	15	96	80
Mo.	St. Louis	− 22	108	31	79	0	98	79
Mont.	Helena	− 42	103	20	66	− 20	90	68

The U factor for the ceiling is found in Table 3-1 to be 0.62. The floor is found to have a U factor of 0.65 and the partition a value of 0.21. Enter these figures in the heat load chart in the proper places (see Figure 3-3).

We can now enter the temperature differences for summer and winter. An indoor temperature of 75°F is generally taken on comfort conditioning applications. Referring to Table 3-2, we note that Dallas is not included in the list. However, we can assume that basically the same conditions exist in Dallas as in Houston. Thus we find a summer design dry bulb temperature of 95°F and a wet bulb temperature of

TABLE 3-2 (Continued)

State	City	Extreme Temperatures		Mean Temperatures		Design conditions		
						Winter	Summer	
		Low	High	January	July	Dry bulb	Dry bulb	Wet bulb
Neb.	Omaha	−32	111	22	77	−15	100	75
Nev.	Winnemucca	−28	104	29	71	−10	95	70
N. C.	Charlotte	−5	103	41	78	10	96	79
N. D.	Bismarck	−45	108	8	70	−25	98	70
N. H.	Concord	−35	102	22	68	−15	95	75
N. J.	Atlantic City	−7	104	32	72	0	95	75
N. M.	Santa Fe	−13	97	29	69	0	92	70
N. Y.	New York City	−14	102	31	74	0	95	77
Ohio	Cincinnati	−17	105	30	75	0	98	77
Okla.	Oklahoma City	−17	108	36	81	0	100	76
Ore.	Portland	−2	104	39	67	10	95	70
Penna.	Philadelphia	−6	106	33	76	0	95	78
R. I.	Providence	−12	101	29	72	0	95	75
S. C.	Charleston	7	104	50	81	15	96	80
S. D.	Pierre	−40	112	16	75	−20	100	72
Tenn.	Nashville	−13	106	39	79	5	98	79
Texas	Galveston	8	101	54	83	25	95	78
Utah	Salt Lake City	−20	105	29	76	−5	95	70
Vt.	Burlington	−28	100	19	70	−15	92	73
Va.	Norfolk	2	105	41	79	10	98	78
Wash.	Seattle	3	98	40	63	10	90	67
W. Va.	Parkersburg	−27	106	32	75	−5	96	77
Wis.	Milwaukee	−25	102	21	70	−15	94	75
Wyo.	Cheyenne	−38	100	26	67	−15	92	70

(Courtesy of American Society of Heating, Refrigerating and Air-Conditioning Engineers, Inc., Atlanta, Ga.)

78°F. We have a design temperature difference of $95 - 75 = 20$. Enter this figure in the TD column, as shown in Figure 3-3.

The problem now is to calculate the area and heat gain for each construction item in each room listed on the heat load chart. We begin with the living room. Referring to Figures 3-1 and 3-3, we can determine the areas involved and make these calculations. We consider the hall as being part of the living room. The area of the hall is 16 ft × 8 ft = 128 sq ft minus the door and window. $D_1 = 19.98$ sq ft, $W_1 = 18$ sq ft. Thus the net wall area = $128 - 19.98 - 18 = 90.02$. We can enter these figures in the proper places on the chart. Note that these particular tables do not differentiate the direction that the windows face. Therefore, we can make this entry anywhere on the chart under the "Windows" heading. When a table is used that makes this differentiation, the areas should be placed in their proper place. The partition has an area of $18 \times 8 = 144$ minus the door area. Thus $144 - 16.65 = 127.35$. Enter this figure in the proper column. We can now use the formula $Q = A \times U \times TD$ to calculate the amount of cooling required for the living room.

The total area of the glass in the living room is 18 sq ft with a U factor of 0.45 and a temperature difference of 20°F. Thus $Q = 18 \times 0.45 \times 20 = 162$ Btu/hr is gained. Enter this figure in the column labeled "Btu/hr Cool." We can calculate and enter the figures for the walls and partitions in the living room. Calculate only those partitions that are not adjacent to a conditioned space. We do not calculate a sun load for the living room because the highest peak load is on the east wall.

Calculate the ceiling, floor, and west door in the same manner as walls, partitions, and glass. Notice on line 22 that we have added an internal cooling load of 2200 in the "Btu/hr Cool" column under "Kitchen" to compensate for cooking, washing dishes, and so on, that occurs in the kitchen area. These should be added in with the other calculations and the total entered on line 23. We find the total Btu/hr cool load on the living room to be 9671.99 Btu. This figure may be rounded off to omit the decimal. However, let us leave it as is until we have calculated the cooling load for all the rooms, added their totals together, and entered them on line 23 under the "Entire House Btu/hr Cool" heading.

Complete the cooling calculations for the remainder of the building as described above and enter the calculations in their proper places. Add the room tables together and enter this figure under the "Entire House Btu/hr Cool" column. Thus we would enter the total from line 23 in line 24. This is the size of cooling unit needed to cool this house under these conditions.

PROCEDURE FOR CALCULATING HEAT LOSS

In presenting this procedure we will calculate the heat loss for a single-story residential building. The type of construction is listed under the heading "Construction Details for Figure 3-1."

While completing this heat loss calculation, refer to the text paragraphs, figures, and tables indicated. The information found there will aid in each step while progressing through the example. The assumed design conditions, orientation, type of construction, and so on, should be indicated on the survey and checklist (see Figure 3-2). Refer to Figure 3-1 for the information required on the survey and checklist.

This example is presented only as an illustration for the use of data and the procedures used in calculating building heat loss. Therefore, it is not to be considered as a recommendation for construction details.

At this point a great amount of the work has already been completed. We have determined the U factors and the areas of the various construction components. We must now determine the temperature differential for the area and design conditions (see Table 3-2). We find that the winter design dry bulb temperature for our area is 25°F. The temperature difference, then, is 75°F (inside temperature) $- 25°F = 50°F$. We can now enter this figure in the "TD Cool/Heat" column, as indicated in Figure 3-6.

It should be noted that we do not use the sun load when calculating the heat loss of a building because the greatest heat loss generally occurs during the night when the sun is not shining. Therefore, the construction components are included in the calculations for their specific rooms at normal conditions.

In calculating the heat loss, we will begin with the living room and compute the heat loss for all the rooms. Then we will place the sum of the room totals under the "Entire House Btu/hr Heat" column when we have completed the calculations.

The living room had no sun load in the heat gain calculations, so there are no areas to be changed. The heat loss for the windows in the living room is: $Q = A \times U \times TD = 18 \times 0.45 \times 50 = 405$ Btu. Enter this figure in the column headed "Living Room Btu/hr Heat" column (see Figure 3-6). Complete the calculations for the wall, partition, ceiling, floor, and door and enter their heat loss on the chart. Add all these figures and enter the sum on line 21 in the heat column. Line 22 indicates that 20% extra should be added. This is required only for rooms with special considerations, such as extra-loose construction or an extra amount of shading. We have no rooms that require this extra percentage. Therefore, enter the figure on lines 21 and 23.

The dining room is next in line to be estimated. We must remember that there was a sun load figured for this room in the heat gain calculation. We must therefore include the wall and glass in their normal place (see Figure 3-6).

Continue to make the calculations for each construction component in each room and enter the figures in their proper place on the heat load chart. Add all the figures in the "Btu/hr Heat" column and enter their totals on lines 21 and 23. Add all the figures on line 21 and enter the sum in the "Btu/hr Heat" column in the "Entire House" column. Also, place this figure on line 23. This is the output capacity of the required heating unit. Although we used a separate load sheet for the heating and cooling calculations, they are normally completed on one (see Figure 3-7).

RESIDENTIAL HEATING AND COOLING LOAD ESTIMATE WORKSHEET

	ORIENTATION	TD COOL/HEAT	U FACTOR	ENTIRE HOUSE	Btu/hr		LIVING ROOM	Btu/hr		DINING ROOM	Btu/hr		KITCHEN	Btu/hr	
				Area	Cool	Heat	Area	Cool	Heat	Area	Cool	Heat	Area	Cool	Heat
WINDOWS															
1	N/S	/50	0.45				18		405	30		675	9		202.5
2															
3	NE/NW														
4															
5	E/W														
6															
7	SE/SW														
8															
WALLS AND PARTITIONS															
9	Wall	/50	0.18				90.02		810.18	30		270	83.8		754.2
10															
11	Part	/50	0.21				127.35		1337.18	104		1092	48		504
12															
SUN LOAD															
13															
14															
15															
16															
FLOOR AND CEILING															
17	Ceiling	/50	0.62				248		7688	88		2728	33.6		1041.6
18	Floor	/50	0.65				248		8060	88		2860	33.6		1092
DOORS															
19	West	/50	0.38				19.98		379.6						
20															
21		Heat loss subtotal				88628.08			18679.88			7625			3594.3
22	External cooling load	20% extra													
23	Cooling load	Heat loss				88628.08			18679.98			7625			3594.3
24	%Total load														
APPROXIMATE AIR QUANTITIES															
25															

FIGURE 3-6
Survey and checklist.

BATH 1			BEDROOM 1			BEDROOM 2			BEDROOM 3			DEN		
	Btu/hr			Btu/hr			Btu/hr			Btu/hr			Btu/hr	
Area	Cool	Heat	Area	Cool	Heat	Area	Cool	Heat	Area	Cool	Heat	Area	Cool	Heat
			27		607.5	12		270	30		675			
			241		2169	104		936	210		1890	76.02		684.18
60		1860	178.5		5533.5	246.5		7641.5	158		4898	162		5022
60		1950	178.5		5801.25	246.5		8011.25	158		5135	162		5265
												19.98		379.62
		3810			14111.25			16858.75			12598			11350.8
		3810			14111.25			16858.25			12598			11350.8

RESIDENTIAL HEATING AND COOLING LOAD ESTIMATE WORKSHEET

	ORIEN-TATION	TD COOL/HEAT	U FACTOR	ENTIRE HOUSE	Btu/hr		LIVING ROOM	Btu/hr		DINING ROOM	Btu/hr		KITCHEN	Btu/hr	
				Area	Cool	Heat	Area	Cool	Heat	Area	Cool	Heat	Area	Cool	Heat
WINDOWS															
1	N/S	20/50	0.45				18	162	405	30		675	9		202.5
2															
3	NE/NW														
4															
5	E/W														
6															
7	SE/SW														
8															
WALLS AND PARTITIONS															
9	Wall	20/50	0.18				90.02	324.07	810.18	30		270	44.8	161.28	754.2
10															
11	Part.	20/50	0.21				127.35	534.87	1337.18	104	436.8	1092	48	201.6	504
12															
SUN LOAD															
13		18/50	0.18							30	97.2		39	126.36	
14		78/50	0.45							30	1053		9	315.9	
15		78/50	0.38												
16															
FLOOR AND CEILING															
17	Ceiling	20/50	0.62				248	3075.2	7688	88	1091.2	2728	33.6	416.64	1041.6
18	Floor	20/50	0.65				248	3224	8060	88	1144	2860	33.6	436.8	1092
DOORS															
19	West	20/50	0.38				19.98	151.85	379.6						
20															
21		Heat loss subtotal				88628.08			18679.98			7625			3594.3
22	Internal cooling load	20% extra						2200						1500	
23	Cooling load	Heat loss		41045.58	88628.08		9671.99	18679.98		3822.2	7625		3158.58	3594.3	
24	%Total load														
APPROXIMATE AIR QUANTITIES															
25															

FIGURE 3-7
Survey and checklist.

	BATH 1			BEDROOM 1			BEDROOM 2			BEDROOM 3			DEN		
		Btu/hr			Btu/hr			Btu/hr			Btu/hr			Btu/hr	
Area	Cool	Heat	Area	Cool	Heat	Area	Cool	Heat	Area	Cool	Heat	Area	Cool	Heat
			27	243	607.5	12	108	270	12	108	675			
			241	867.6	2169	104	374.4	936	68	244.8	1890	76.02		684.18
												16.65	69.93	
									142	460.08		76.02	246.3	
									18	631.8				
												19.98	592.2	
60	744	1860	178.5	2213.4	5533.5	246.5	3056.6	7641.5	158	1959.2	4898	162	2008.8	5022
60	780	1950	178.5	2320.5	5801.25	246.5	3204.5	8011.25	158	2054	5135	162	2106	5265
												19.98		379.62
		3810			14111.25			6858.75			12598			11350.8
	1524	3810		5644.5	14111.25		6743.5	16858.75		5457.88	12598		5022.9	11350.8

According to our calculations we need a cooling unit with a capacity of 41,046 Btu/hr and a heating unit with an output (bonnet) capacity of 76,220 Btu. In Chapter 4 we will choose the equipment that fits our needs most closely. In Chapter 7 we will size a duct system to match our unit selections.

SUMMARY

The size of air conditioning and heating equipment is calculated by the actual peak load requirements.

The peak sun load will occur, in most cases, at one time. However, if an equal load occurred at more than one time, it would make no difference which time was chosen, especially for residential applications.

The peak sun load may occur at a time when none of the construction items are at a peak.

The sun load is not considered in the heating calculations because the greatest heating load occurs at night.

When obtaining the design conditions from a chart, use a town or city near your locality if your particular locality is not listed.

When calculating heat loads, use only the net areas of walls.

Each room of a residence should be considered individually, then the sum of the heat loads entered under the "Entire House" heading to indicate the required equipment size.

An internal cooling load of 2200 Btu/hr has been added to the living room calculation to compensate for the people and 1500 Btu/hr has been added to the kitchen to compensate for cooking, washing dishes, and so on.

REVIEW QUESTIONS

1. Determine the quantity of infiltration air that should be used in calculating the infiltration heat load for a room having the following windows: three windows 3 ft × 5 ft in one wall and two windows 2 ft × 5 ft in the opposite wall, all of wood, double hung, of average fit, and not weather-stripped. The third wall has two good-fitting casement windows 18 in. × 36 in. Assume a 20-mph wind velocity.

2. Calculate the infiltration heat load gained from 1750 cu ft of air per hour in a summer air conditioning installation located in Indianapolis, Indiana. Use an inside design condition of 80°F dry bulb and 50% relative humidity. Show total heat, latent heat, and sensible heat for this problem.

3. How much differential should be figured for a partition next to a kitchen when the normal differential for the exterior wall is 20°F?

4. A 2-hp electric motor gives off its heat in the conditioned space. How many Btu is given off by the motor?

5. Determine the maximum sun effect through 100 sq ft of bare glass in a west wall.

6. A basement restaurant has twenty 100-W light bulbs that are used continually. The electric appliances total 10,000 W, but they are used only during one-fourth of the peak period. What total heat load is placed on the air conditioning plant by the lights and the appliances?

7. Determine the conduction heat load through a wall 30 ft long and 8 ft high. The wall has four double-glass windows 4 ft × 5 ft. The wall is constructed of wood siding, 2 in. × 4 in. studs, plastered. Use a 15°F differential.

8. What is the least amount of outside air required in a beauty parlor having a maximum of 10 customers and 4 employees?

9. What would be the sensible, latent, and total heat given off by 10 patrons and 4 employees in a beauty parlor?

10. What is the temperature differential for excess sun load for the following?
    a. East wall, medium color value, at 11:00 A.M.
    b. Light-colored roof at 12:00 noon
    c. West wall, light color, at 5:00 P.M.
    d. East glass, bare, at 7:00 A.M.
    e. South glass, awnings, at 3:00 P.M.

4 Equipment Sizing and Selection

The proper sizing and selection of air conditioning equipment is just as important as any other step in the estimating process. When the equipment is sized too small, the temperature inside the building will not be properly maintained on warmer days. Similarly, if the equipment is sized too large, humidity control and operating costs will be poor.

INTRODUCTION

The proper sizing and selection of air conditioning equipment can be accomplished only after an accurate heat loss and heat gain analysis is completed, as described in Chapters 2 and 3. The size of the equipment is obtained from the load calculation form, and manufacturers' tables and charts will provide the selection of equipment that will fulfill our needs most exactly.

The necessity for proper sizing of equipment cannot be overemphasized. A unit that is sized too small will cost less to install but will operate continuously with poor temperature control and customer dissatisfaction. Continuous operation will also shorten the life of the unit. Too large a unit will increase the initial cost of the unit with poor humidity control. The operating expense will also increase with a unit that is too large, also resulting in customer dissatisfaction.

HEATING EQUIPMENT SIZING

The proper balance of equipment and load is just as important as balancing the compressor and evaporator, balancing the vapor velocity and the pressure drop in refrigeration piping, or any other component balancing. Not only is the balance between different components of the system vital, but the balance between system capacity and load, the balance between cost and performance, and many other factors must be given consideration.

With the almost infinite number of design and operating conditions possible, equipment that meets all the requirements exactly is seldom available. Thus the selection of equipment becomes a problem of choosing the unit that is the closest to satisfying the demands of capacity, performance, and cost. A closely related problem is that of determining which considerations can, and which cannot, be compromised. Also important are learning how equipment capacity is determined and the performance characteristics of the various models, sizes, and combinations of components.

It should be noted, however, that most of this work has already been done by the manufacturer and its engineering staff. These people have completely analyzed the problems and have compiled their answers in charts, tables, and formulas which are all easy to use, thus making selection a matter of understanding and using all the selection aids that are available (see Figure 4-1). In our discussion we start by choosing units that are relatively simple to select, then gradually move on to the more difficult selections.

GENERAL INFORMATION

The selection of furnaces and other packaged units is relatively simple because the system components have already been chosen by the equipment manufacturer. Therefore, only the balancing of the unit capacity to the heating load and the determination of such items as the cubic feet per minute (cfm) delivered by the unit and the required electrical characteristics are left.

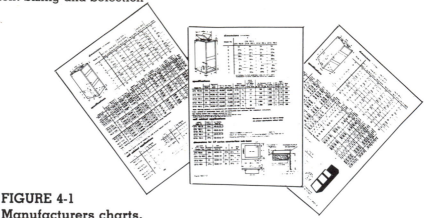

FIGURE 4-1
Manufacturers charts.

■ Data Required for Heating Unit Selection

There is certain information that must be determined before attempting to select the equipment. The following information must be known before attempting to select warm air furnaces for residential or light commercial installations.

1. The inside and outside design temperatures
2. The total heat loss in Btu/hr
3. The type of heating equipment to be used
4. The required cfm
5. The external static pressure caused by the ductwork

The Inside and Outside Design Temperatures. When considering these temperatures, only the dry bulb temperatures are required. As stated earlier, inside design temperatures are usually between 70 and 75°F (21.11 and 23.89°C).

The Total Heat Loss in Btu/hr. The total heat loss figure is determined by completing a heat loss form. When it is necessary to consider the heat loss from ducts that are installed in unconditioned spaces, there are two ways to determine the figure: (1) use the method included on the heat loss form, and (2) multiply the calculated heat loss by 0.95; thus the actual amount of heat that is delivered to the building is 95% of the output capacity of the furnace selected.

Type of Heating Equipment. The type of furnace selected will depend on several application variations, such as the type of energy available, or desired, and whether an upflow, downflow, or horizontal airflow is desired.

FIGURE 4-2
**Upflow furnace. (Courtesy of Southwest Manufac-
turing Co.)**

The type of energy used—gas, oil, or electricity—will depend to a
great deal on the availability and economy of each source. The convenience and dependability of each heat source should also be considered.
In rural areas in which natural gas will be available in the near future,
it may be wise to suggest that a liquefied petroleum (LP) gas furnace
be installed, then converted when natural gas becomes available.

The upflow forced warm air furance is the model most commonly
used in residential installations (see Figure 4-2). When the upflow furnace is in operation, air is drawn in through the bottom of the unit, is
forced through and around the heat exchanger, and is discharged out
the top of the unit. These types of furnaces are ideally used in basement installations with under-the-floor air distribution systems in
conjunction with floor-level or high-wall diffusers. They are also excellent for use in ground floor installations connected to an air distribution system in the attic using overhead or high-wall diffusers.

Downflow, or counterflow, furnaces have the air handling section
and the burner-heat exchanger sections reversed (see Figure 4-3). In
operation, the air is drawn in the top of the furnace, is forced through
and around the heat exchanger, and is discharged out the bottom of
the unit. These types of units are ideal for basement-less homes or
commercial buildings that are built either on a slab or over a crawl
space. The furnace is installed over a hole in the floor. The discharge
air plenum and the duct system are installed below the floor (see Figure 4-4). Floor-type diffusers are usually used with this type of
system.

FIGURE 4-3
Downflow furnace. (Courtesy of Southwest Manu-
facturing Co.)

Horizontal furnaces are ideal for installations where floor space is limited or valuable. Usually, this type of furnace is installed in the attic, in a crawl space, or may be suspended from the ceiling (see Figure 4-5). In operation, the air is drawn in through one end of the unit, forced through and around the heat exchanger, then forced out the

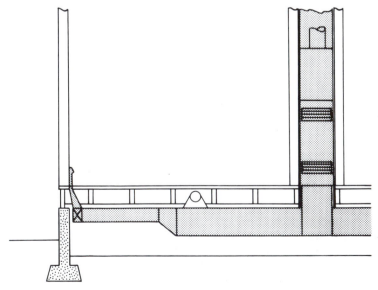

FIGURE 4-4
Downflow installation.

FIGURE 4-5
Horizontal furnace. (Courtesy of Southwest Manufacturing Co.)

other end. The air then flows through the duct system to the heated space (see Figure 4-6).

The direction of airflow and the type of energy chosen must be known before the furnace manufacturer's tables can be used. When the furnace is to be used as a part of a year-round system, the type and size of the recommended cooling equipment might influence the choice of heating unit.

The Required cfm.   To properly distribute the heated air in the conditioned space will require that the cfm be properly calculated before

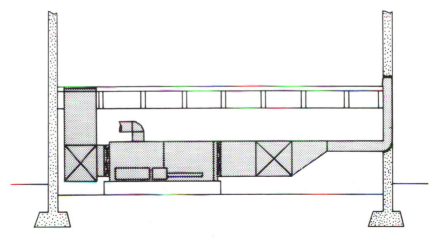

FIGURE 4-6
Horizontal installation.

the duct system is designed. There are various methods of duct design for residential and small commercial applications presented by various organizations, such as the Natural Warm Air Heating and Air Conditioning Association and ASHRAE. The method used will depend on personal preference.

External Static Pressure Caused by the Ductwork.   The external static pressure at the required cfm is a vital part of the complete duct system design calculations. It should therefore be included in the duct system specifications.

■ Furnace Selection Procedure

When the total heat loss has been calculated, there are only two steps left in selecting a furnace for a heating-only system. They are (1) selecting the furnace from manufacturers' tables, and (2) determining the necessary cfm adjustment for heating.

In our selection we will use the building specifications below for heat loss and heat gain calculations (see Figure 4-7).

Use the following information and steps to select a gas-fired upflow furnace to be used in a building of the description following.

■ Construction Details for Figure 4-7

- Walls: Brick veneer, 2 in. × 4 in. studs, plastered, ½-in. batt insulation
- Windows: All are double glass.
        Window sizes are: $W_1 = 3$ ft × 6 ft, $W_2 = 3$ ft × 5 ft, $W_3 = 3$ ft × 4 ft, $W_4 = 3$ ft × 3 ft
- Doors: Wood with storm doors and weather stripped.
        Door sizes: $D_1 = 3$ ft 0 in. × 6 ft 8 in., $D_2 = 2.5$ ft × 6 ft. 8 in.
- Ceiling: Plastered without flooring above.
- Floor: 4 in. concrete slab.
- Warm Partition: Basic frame construction with ½ in. of insulation.
- Ducts: Located in attic with flexible insulation. We will assume that the duct system will be designed to provide an external static pressure of 0.20 in. (0.508 mm) water column pressure.

The house is located in Dallas, Texas.

Selection.   We determined that the total heat loss for the building was 76,220 Btu/hr (line 24 under the "Entire House" column). This figure includes the heat loss of the structure and the duct heat loss. Thus the furnace we select must have an output (bonnet) capacity of at least 80,000 Btu/hr.

1. Some manufacturers' capacity data tables list output capacities, whereas others list input capacities. When the output capacity is listed, this figure may be used for selecting the furnace. If the input capacity is listed, the output capacity can be determined by multiplying the rated capacity by 0.80. This is because the American Gas Association requires that all AGA-approved gas-fired units be rated at 80% efficiency. Example: A 100,000-Btu/hr input gas-fired furnace will have an output capacity of $100,000 \times 0.80 = 80,000$ Btu/hr.

2. Next, we select a furnace that has a capacity equal to or greater than 76,220 Btu/hr. Figure 4-8 shows a manufacturer's specifications chart for an upflow gas-fired furnace. Note that furnace model number HL100-D is the nearest to our required heating capacity. It has a Btu input of 100,000 Btuh/hr for elevations up to 2000 ft and an output of 80,000 Btuh/hr for the same altitudes.

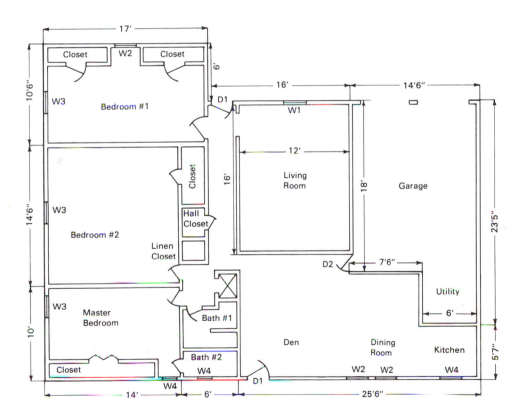

FIGURE 4-7
Model house.

3. This furnace has an air-handling capacity of 925 cfm with a 75°F temperature rise through the furnace. It will also deliver 1000 cfm for air conditioning applications with an external static pressure on the duct system of 0.50 in. water column. It has an AGA listed external static pressure of 0.30 in. water column at a 45 to 75°F temperature rise through the furnace. This is the furnace to use.

4. The furnace is 51 in. high and 18½ in. wide and is 27¼ in. deep. (see dimensions A, B, and C in Figure 4-9). If a furnace of these dimensions will fit into the designated space, the installation will be simple. However, if the opening is smaller than the furnace, the opening must be increased to accommodate the furnace, or the furnace must be installed in another location.

5. The required vent diamenter is 5 in. This must be taken into account when figuring the vent cost because an error here would be very costly. The double wall vent pipe is very expensive. Also, if replacing a furnace, the existing venting system may be usable and the proper vent pipe transitions must be determined.

6. The furnace weighs 190 lb. Thus the flooring must be strong enough to support at least this much weight. The weight of the duct work should also be added to this figure when determining the required strength of the supporting platform.

Model No.	Cabinet Width	Optional Return Air Drop	*B.T.U. Per Hour Input	Output	Blower Wheel Size	Motor H.P.	A.G.A. Listed Ext. Static Pr. @ 45°-75° Temp. Rise	Air Delivery, C.F.M. Heating @ 75° T.R.	†Air Cond. @ .50 E.S.P.	A.C. Cap. Tons	Vent Dia.	Permanent Filter 1" Thick (Furnished)	Ship Wt.
HL50-D	15¼	HRD-1	50,000	40,000	DD9-6	¼	.30"	465	700	1¾	4	12x25	140
HL75-DD	15¼	HRD-1	75,000	60,000	DD9-6	¼	.30"	695	800	2	4	12x25	152
HL80-D2	15¼	HRD-1	80,000	64,000	DD9-7A	¼	.40"	740	800	2	4	12x25	152
HL75-B3	18½	HRD-1	75,000	60,000	A12-9	⅓	.30"	695	1,200	3	4	16x25	176
HL100-D	18½	HRD-1	100,000	80,000	DD9-9	¼	.30"	925	1,000	2½	5	16x25	19⁻
HL100-B3	18½	HRD-1	100,000	80,000	A12-9	⅓	.30"	925	1,200	3	5	16x25	196
HL100-B4	23	HRD-2	100,000	80,000	A12-9	½	.30"	925	1,600	4	5	20x25	205
HL125-B3	23	HRD-2	125,000	100,000	A12-12	⅓	.40"	1,160	1,200	3	6	20x25	221
HL125-B4	23	HRD-2	125,000	100,000	A12-12	½	.40"	1,160	1,600	4	6	20x25	228
HL125-B5	27½	HRD-2	125,000	100,000	A12-12	¾	.40"	1,160	2,000	5	6	25x25	248
HL150-B4	27½	HRD-2	150,000	120,000	A12-15	½	.40"	1,390	1,600	4	7	25x25	252
HL150-B5	31	HRD-2	150,000	120,000	A12-15	¾	.40"	1,390	2,000	5	7	(2) 16x25	282
HL150-B75	31	HRD-2	150,000	120,000	A12-15	1	.40"	1,390	3,000	7½	7	(2) 16x25	290

* Ratings shown are for elevations up to 2,000 feet. For elevations above 2,000 feet, ratings should be reduced at the rate of 4 percent for each 1,000 feet above sea level.

†Gas heating section only is Design certified by A.G.A.

Filter Rack built in on all models for side and/or bottom return air.

FIGURE 4-8

Furnace specifications chart. (Courtesy of Southwest Manufacturing Co.)

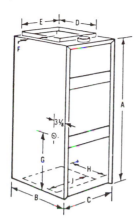

dimensions (in inches)

Model No.	HL50-D HL75-DD HL80-D2	HL75-B3 HL100-D HL100-B3	HL100-B4 HL125-B3 HL125-B4	HL125-B5 HL150-B4	HL150-B5 HL150-B75
A	51	51	51	51	51
B	27¼	27¼	27¼	27¼	27¼
C	15¼	18½	23	27½	31
D	19⅝	19⅝	19⅝	19⅝	19⅝
E	13¼	16½	21	25½	29
F	1	1	1	1	1
G	28	28	28	28	28
H	23¼	23¼	23¼	23¼	23¼
J	10¾	14	18½	23	23

FIGURE 4-9
Furnace dimensions chart. (Courtesy of South-
west Manufacturing Co.)

COOLING EQUIPMENT SIZING

As stated earlier under the heading "Heating Equipment Sizing," the
proper balance of equipment and load is very important. Also, the balancing of system components, such as the condensing unit and the
evaporator, and the air delivery and the duct system, is important if
comfort is to be maintained at maximum efficiency.

GENERAL INFORMATION

The selection of cooling equipment, whether it be a remote, split system or a packaged unit, has been made relatively simple because the
equipment manufacturers have already determined what components
will produce a given unit capacity. Therefore, only the balancing of
the unit capacity to the cooling load, and the determination of such
items as the cfm delivered by the unit and the required electrical characteristics are left.

■ Data Required for Cooling Unit Selection

There is certain information that must be determined before attempting to select the equipment. The following information must be known
before the proper selection of cooling equipment for residential or
light commercial installations can be completed:

1. The summer design outdoor air dry bulb (DB) and wet bulb (WB) temperatures

2. The conditioned space design DB and WB temperatures

3. The required ventilation air in cfm

4. The total cooling load in Btu/hr

5. The cfm air delivery required

6. The condensing medium (air or water) and the temperature of the medium entering the condenser

7. The duct external static pressure

The Design Outdoor Air Temperatures. The design outdoor air conditions are necessary for two reasons: (1) to complete the cooling (heat gain) load estimate, and (2) to establish the conditions of the air that enters the evaporator coil when positive ventilation is provided.

The design outdoor air temperatures used should be those commonly used in the geographical area where the building is located. The design outdoor air temperatures for most large U.S. cities can be found in tables (see Table 3-2). If the area under consideration is not listed in the table, use the closest city within 100 miles. When no listing is shown, within a suitable distance, the accepted outdoor design temperatures can be obtained from a heating, ventilating, and air conditioning guide, a local consulting engineer, or a local weather bureau.

The Conditioned Space Design DB and WB Temperatures. The conditioned space design temperatures are also necessary for two reasons: (1) the load calculation, and (2) the evaporator entering air temperature calculation. The air entering the evaporator is a mixture of outside air and the air returned from the room to the evaporator. Thus, when the temperatures of the return air (indoor design) and the ventilation air (outdoor design), together with the percentage of ventilation air, are known, the temperature of the air entering the evaporator can be easily calculated.

When considering bid-specification jobs, the indoor air conditions will be specified by the engineer. When other types of jobs are considered, the design indoor conditions are established which are acceptable to the owner and which are consistent with local experience and recommendations by the industry.

The Required Ventilation Air in cfm. The cfm of ventilation air required must be known before attempting to determine the temperature of the air entering the evaporator or attempting to determine the sensible heat capacity of the air conditioning system. The ventilation air requirement is determined from the cooling load calculation and should be obtained from the load calculation sheet.

The Total Cooling Load in Btu/hr.   The total cooling load (heat gain) is the basis for the complete air conditioning unit selection. The tentative and final equipment selection is made on the basis of this calculation.

The heat gain load is found on the heat gain calculation form. There are several short forms available for making this estimate. However, should special conditions be present, or if the construction features are considerably different from those described on the calculation form, a more detailed load estimating method should be used.

The cfm Air Delivery Required.   When the long-form heat gain calculation form is used, the cfm required can be calculated from the total sensible heat load and the rise in the DB temperature of the air supply. On some load calculation forms no cfm requirement is specified. In such cases, the unit is chosen on the Btu/hr capacity alone. It is then assumed that the rated air delivery is 400 cfm per ton of refrigeration. This figure of 400 cfm is used in making duct static pressure calculations.

The Condensing Medium (Air or Water) and the Temperature of the Medium Entering the Condenser.   When water-cooled condensers are used with well water or city water, the entering water temperature must be determined to calculate properly the gallons per minute (gpm) of water that must flow through the condenser to achieve the design condensing temperatures. When a cooling tower is used, the discharge water temperature at the tower must be known to calculate the gpm flow and the final condensing temperature.

The city or well water temperature can be measured at the source of supply. The cooling tower condenser water temperatures are obtained from tower manufacturers' literature (see Table 4-1).

The capacity of an air-cooled condenser is determined on the difference between the condensing temperatures and the design outdoor air DB temperatures. Most equipment manufacturers publish these data in tables (see Table 4-2). The entering air temperatures for air-cooled condensers are the various design air temperatures.

The Duct External Static Pressure.   When an air conditioning unit is to be connected to a duct system, the external static pressure must be known to determine the fan speed and fan motor horsepower required. The maximum external static pressure for the equipment is determined by the manufacturer and is indicated on the equipment nameplate. The external static pressure for the duct system is calculated

(Courtesy of The Marley Company)

TABLE 4-1
Cooling Tower Rating Chart

No.	Capacity Tons-Refrig. 78° WB	Overall Dimensions Wide	Deep	High	Pump Head In Feet	H.P.	Motor Voltage	Cold Water Outlet	Ship. Wt.	Max. Oper. Wt.	FOB Louisville Ea.	Basin Cover Part No.	Basin Cover Ea.	Air Inlet Screen Part No.	Air Inlet Screen Ea.
91000	5	$32\frac{1}{2}''$	$69''$	$60\frac{5}{16}''$	4.5	$\frac{1}{4}$	115/230/60/1	2" FIPT	460 Lbs.	960 Lbs.	$1066.00			91055	$ 38.25
91001	7	$32\frac{1}{2}''$	$69''$	$60\frac{5}{16}''$	4.5	$\frac{1}{3}$	115/230/60/1	2" FIPT	460 Lbs.	960 Lbs.	1079.00			91055	38.25
91002	10	$32\frac{1}{2}''$	$69''$	$61\frac{1}{8}''$	4.5	$\frac{1}{3}$	115/230/60/1	2" FIPT	480 Lbs.	1060 Lbs.	1158.00			91056	40.75
91003	15	$32\frac{1}{2}''$	$69''$	$70\frac{5}{16}''$	5.3	$\frac{1}{2}$	115/230/60/1	2" FIPT	520 Lbs.	1100 Lbs.	1342.00				
91004	15	$32\frac{1}{2}''$	$69''$	$70\frac{5}{16}''$	5.3	$\frac{1}{2}$	200/60/3	2" FIPT	520 Lbs.	1100 Lbs.	1342.00	91050	12.00	91057	52.75
91005	15	$32\frac{1}{2}''$	$69''$	$70\frac{5}{16}''$	5.3	$\frac{1}{2}$	230/460/60/3	2" FIPT	520 Lbs.	1100 Lbs.	1342.00				
91006	20	$32\frac{1}{2}''$	$72\frac{3}{4}''$	$80\frac{1}{2}''$	6.5	$\frac{1}{2}$	115/230/60/1	4" IPT	580 Lbs.	1020 Lbs.	1500.00				
91007	20	$32\frac{1}{2}''$	$72\frac{3}{4}''$	$80\frac{1}{2}''$	6.5	$\frac{1}{2}$	200/60/3	4" IPT	580 Lbs.	1020 Lbs.	1500.00			91058	58.00
91008	20	$32\frac{1}{2}''$	$72\frac{3}{4}''$	$80\frac{1}{2}''$	6.5	$\frac{1}{2}$	230/460/60/3	4" IPT	580 Lbs.	1020 Lbs.	1500.00				
91009	25	$46\frac{1}{2}''$	$75\frac{5}{8}''$	$80\frac{1}{2}''$	6.5	$\frac{1}{2}$	115/230/60/1	4" IPT	750 Lbs.	1410 Lbs.	1645.00				
91010	25	$46\frac{1}{2}''$	$75\frac{5}{8}''$	$80\frac{1}{2}''$	6.5	$\frac{1}{2}$	200/60/3	4" IPT	750 Lbs.	1410 Lbs.	1645.00				
91011	25	$46\frac{1}{2}''$	$75\frac{5}{8}''$	$80\frac{1}{2}''$	6.5	$\frac{1}{2}$	230/460/60/3	4" IPT	750 Lbs.	1410 Lbs.	1645.00				
91012	30	$46\frac{1}{2}''$	$75\frac{5}{8}''$	$80\frac{1}{2}''$	6.5	1	115/230/60/1	4" IPT	750 Lbs.	1410 Lbs.	1776.00				
91013	30	$46\frac{1}{2}''$	$75\frac{5}{8}''$	$80\frac{1}{2}''$	6.5	1	200/60/3	4" IPT	750 Lbs.	1410 Lbs.	1776.00	91051	19.75	91059	72.25
91014	30	$46\frac{1}{2}''$	$75\frac{5}{8}''$	$80\frac{1}{2}''$	6.5	1	230/460/60/3	4" IPT	750 Lbs.	1410 Lbs.	1776.00				
91016	40	$46\frac{1}{2}''$	$75\frac{5}{8}''$	$80\frac{1}{2}''$	8.5	2	200/60/3	4" IPT	750 Lbs.	1410 Lbs.	2053.00				
91017	40	$46\frac{1}{2}''$	$75\frac{5}{8}''$	$80\frac{1}{2}''$	8.5	2	230/460/60/3	4" IPT	750 Lbs.	1410 Lbs.	2053.00				
91018	55	$64''$	$83\frac{1}{16}''$	$91\frac{3}{8}''$	9.0	2	200/60/3	4" IPT	1170 Lbs.	2210 Lbs.	2539.00				
91019	55	$64''$	$83\frac{1}{16}''$	$91\frac{3}{8}''$	9.0	2	230/460/60/3	4" IPT	1170 Lbs.	2210 Lbs.	2539.00	91052	27.75	91060	108.00
91020	65	$64''$	$83\frac{1}{16}''$	$91\frac{3}{8}''$	8.0	3	200/60/3	4" IPT	1190 Lbs.	2230 Lbs.	2829.00				
91021	65	$64''$	$83\frac{1}{16}''$	$91\frac{3}{8}''$	8.0	3	230/460/60/3	4" IPT	1190 Lbs.	2230 Lbs.	2829.00				
91022	75	$76''$	$89\frac{1}{16}''$	$91\frac{3}{8}''$	8.0	3	200/60/3	6" IPT	1400 Lbs.	2770 Lbs.	3158.00	91053	30.25	91061	117.00
91023	75	$76''$	$89\frac{1}{16}''$	$91\frac{3}{8}''$	8.0	3	230/460/60/3	6" IPT	1400 Lbs.	2770 Lbs.	3158.00	91053	30.25	91061	117.00
91024	90	$88''$	$92\frac{3}{4}''$	$91\frac{3}{8}''$	8.0	5	200/60/3	6" IPT	1610 Lbs.	3250 Lbs.	3671.00	91054	33.00	91062	138.00
91025	90	$88''$	$92\frac{3}{4}''$	$91\frac{3}{8}''$	8.0	5	230/460/60/3	6" IPT	1610 Lbs.	3250 Lbs.	3671.00	91054	33.00	91062	138.00

TABLE 4-2
Condensing Unit Sizing Chart

MODEL NO.	COOLING CAPACITIES @ 95° Amp.	EVAPORATOR COIL MODEL NO.	RATED EVAPORATOR AIRFLOW	TOTAL WATTS @ 95°F.	ELECTRICAL 60 Hz. Volts/Ph.	DELAY FUSE SIZE	MIN. WIRE SIZE	DIMENSIONS			APROX. SHIP ST.
								H	W	D	
CTD-201-QC*	24,000	UCC-25-15-QC	800	3,200	230/1	30A	#12	22¼	25¾	31¾	162
CTD-251-QC	29,000	UCC-30-18-QC	1,000	3,600	230/1	40A	#10	22¼	25¾	31¾	172
CTD-301-QC	36,000	UCC-36-18-QC	1,220	4,600	230/1	40A	#10	24¾	25¾	31¾	201
CTD-303-QC	36,000	UCC-36-18-QC	1,220	4,600	208-240/3	30A	#12	24¾	25¾	31¾	201
CTD-351-QC	41,000	UCC-41-23-QC	1,300	5,300	230/1	50A	# 8	24¾	25¾	31¾	208
CTD-353-QC	41,000	UCC-41-23-QC	1,300	5,300	208-240/3	40A	#10	24¾	25¾	31¾	209
CTD-401-QC	47,000	UCC-48-23-QC	1,600	6,200	230/1	50A	# 8	29¾	28	35½	252
CTD-403-QC	47,000	UCC-48-23-QC	1,600	6,200	208-240/3	40A	#10	29¾	28	35½	253
CTD-501-QC	58,000	UCC-61-27-QC	2,000	7,600	230/1	60A	# 6	29¾	28	35½	267
CTD-503-QC	58,000	UCC-61-27-QC	2,000	7,600	208-240/3	50A	# 8	29¾	28	35½	267
CTD-201	24,000	UC-25-18	800	3,300	230/1	30A	#12	22¼	25¾	31¾	172
CTD-251	28,000	UC-25-18	1,000	3,600	230/1	40A	#10	22¼	25¾	31¾	182
CTD-253	28,000	UC-25-18	1,000	3,600	208/240/3	20A	#12	22¼	25¾	31¾	182
CTD-301	36,000	UC-36-18	1,220	4,600	230/1	40A	#10	24¾	25¾	31¾	211
CTD-303	36,000	UC-36-18	1,220	4,620	208-240/3	30A	#12	24¾	25¾	31¾	201
CTD-351	41,000	UC-36-23	1,300	5,300	230/1	50A	# 8	24¾	25¾	31¾	218
CTD-353	41,000	UC-36-23	1,300	5,300	208-240/3	40A	#10	24¾	25¾	31¾	219
CTD-401	47,000	UC-48-23	1,600	6,200	230/1	50A	# 8	29¾	28	35½	257
CTD-403	47,000	UC-48-23	1,600	6,200	208-240/3	40A	#10	29¾	28	35½	258
CTD-501	58,000	UC-61-27	2,000	7,600	230/1	60A	# 6	29¾	28	35½	272
CTD-503	58,000	UC-61-27	2,000	7,600	208-240/3	50A	# 8	29¾	28	35½	272

* "QC" after model number indicates "Quick-Connect" condenser unit and matching evaporator coil.

(Courtesy of Southwest Manufacturing Co.)

TABLE 4-3
Resulting Temperatures of a Mixture of Return Air and Outdoor Air (to be used only for return air having an 80°F DB temperature and 50% humidity)

| Percentage of outside air | Dry Bulb Temperature of the Air Mixture | | | Wet Bulb Temperature of the Air Mixture | | | | | | | Percentage of Outside Air |
| | Outside Dry Bulb Temp. | | | Outside Wet Bulb Temperature | | | | | | | |
	95°	100°	105°	74°	75°	76°	77°	78°	79°	80°	
2%	80.3	80.4	80.5	67.2	67.2	67.2	67.2	67.3	67.3	67.3	2%
4	80.6	80.8	81.0	67.3	67.3	67.4	67.4	67.5	67.5	67.6	4
6	80.9	81.2	81.5	67.5	67.5	67.6	67.7	67.7	67.8	67.9	6
8	81.2	81.6	82.0	67.6	67.7	67.8	67.9	68.0	68.1	68.2	8
10	81.5	82.0	82.5	67.8	67.9	68.0	68.1	68.2	68.4	68.5	10
12%	81.8	82.4	83.0	67.9	68.0	68.2	68.3	68.5	68.6	68.8	12%
14	82.1	82.8	83.5	68.1	68.2	68.4	68.5	68.7	68.9	69.1	14
16	82.4	83.2	84.0	68.2	68.4	68.6	68.8	69.0	69.2	69.4	16
18	82.7	83.6	84.5	68.3	68.5	68.8	69.0	69.2	69.4	69.6	18
20	83.0	84.0	85.0	68.5	68.7	69.0	69.2	69.4	69.7	69.9	20
22%	83.3	84.4	85.5	68.6	68.9	69.1	69.4	69.7	69.9	70.2	22%
24	83.6	84.8	86.0	68.8	69.1	69.3	69.6	69.9	70.2	70.5	24
26	83.9	85.2	86.5	68.9	69.2	69.5	69.8	70.1	70.4	70.8	26
28	84.2	85.0	87.0	69.1	69.4	69.7	70.0	70.4	70.7	71.0	28
30	84.5	86.0	87.5	69.2	69.6	69.9	70.2	70.6	70.9	71.3	30

when the ductwork is designed and should be included in the duct system specifications. In any case it should never exceed the manufacturer's specifications.

■ Split-System Equipment Selection Procedure

Basically, the same selection procedure is used for both residential and commercial split-system equipment applications. Before the actual equipment selection can be done, the heat gain calculations must be completed. The equipment can then be chosen according to the appropriate following application and installation considerations:

1. The temperatures of the air entering the evaporator and condenser are determined. The design outdoor air DB temperature is the condenser entering air temperature. In residential applications and other types where no ventilation air is brought in through the equipment, the design conditioned space DB and WB temperatures are used as the evaporator entering air temperatures. When the ventilation air is introduced through the evaporator, the entering air temperatures may be calculated or obtained from a table (see Table 4-3).

2. Next a split system with sufficient capacity is selected from the manufacturers' tables. When sizing commercial units, the total capacity should be equal to or greater than the total heat gain load, and the sensible capacity should be enough to take care of the sensible heat load. When sizing equipment for residential applications, the unit capacity can be a little less than the calculated heat gain and should be only a very small amount more.

3. The air delivery by the unit against the calculated external static pressure must be determined. When the cooling cfm requirement is considerably more than the heating requirement, the furnace may be equipped with a two-speed fan motor.

For an example of the selection of a split-system air conditioning unit, let us use the residence that was used in the example for selecting a gas-fired upflow furnace earlier in the chapter. Let us assume that a year-around air conditioning unit is to be installed. The heat gain calculation was completed in Chapter 3. We will use the same construction details that were used in the furnace selection.

Selection. We determined that the total heat gain for our building was 41,045.58 (line 24 under the "Entire House" column). This figure includes the sensible heat gain of the structure, the duct heat gain, and the latent heat gain. Thus the cooling unit selected must have a total Btu/hr capacity of 41,045.58 Btu/hr.

The heating unit must have an output (bonnet) capacity of at least 76,220 Btu/hr.

1. Some manufacturers' data lists the output capacities at different outdoor ambient temperatures, whereas others just list them at one temperature (see Table 4-2).

2. Next, we select a condensing unit that has a capacity equal to or slightly larger than 41,045 Btu/hr. Using Figure 4-10, locate the heading "Performance A.R.I. Cap." We find that condensing unit model CTD-351-QC meets our requirements. It has a capacity of 41,000 Btu (3½ tons approximately) per hour with matching coils model UCC-41-23-QC and 1300 cfm over the evaporator coil.

3. We must match the coil to the furnace air outlet opening dimensions as closely as possible. The evaporator coil configuration and dimensions are shown in Figure 4-10. The furnace dimensions are

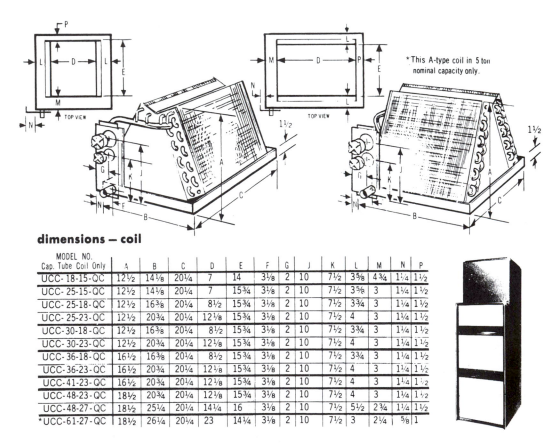

dimensions — coil

MODEL NO. Cap. Tube Coil Only	A	B	C	D	E	F	G	J	K	L	M	N	P
UCC-18-15-QC	12½	14⅛	20¼	7	14	3⅛	2	10	7½	3⅝	4¾	1¼	1½
UCC-25-15-QC	12½	14⅛	20¼	7	15¾	3⅛	2	10	7½	3⅝	3	1¼	1½
UCC-25-18-QC	12½	16⅜	20¼	8½	15¾	3⅛	2	10	7½	3¾	3	1¼	1½
UCC-25-23-QC	12½	20¾	20¼	12⅛	15¾	3⅛	2	10	7½	4	3	1¼	1½
UCC-30-18-QC	12½	16⅜	20¼	8½	15¾	3⅛	2	10	7½	3¾	3	1¼	1½
UCC-30-23-QC	12½	20¾	20¼	12⅛	15¾	3⅛	2	10	7½	4	3	1¼	1½
UCC-36-18-QC	16½	16⅜	20¼	8½	15¾	3⅛	2	10	7½	3¾	3	1¼	1½
UCC-36-23-QC	16½	20¾	20¼	12⅛	15¾	3⅛	2	10	7½	4	3	1¼	1½
UCC-41-23-QC	16½	20¾	20¼	12⅛	15¾	3⅛	2	10	7½	4	3	1¼	1½
UCC-48-23-QC	18½	20¾	20¼	12⅛	15¾	3⅛	2	10	7½	4	3	1¼	1½
UCC-48-27-QC	18½	25¼	20¼	14¼	16	3⅛	2	10	7½	5½	2¾	1¼	1½
*UCC-61-27-QC	18½	26¼	20¼	23	14¼	3⅛	2	10	7½	3	2¼	⅝	1

FIGURE 4-10
Evaporator dimensions. (Courtesy of Southwest Manufacturing Co.)

(in inches) Model No.	HL50-D HL75-DD HL80-D2	HL75-B3 HL100-D HL100-B3	HL100-B4 HL125-B3 HL125-B4	HL125-B5 HL150-B4	HL150-B5 HL150-B75
A	51	51	51	51	51
B	27¼	27¼	27¼	27¼	27¼
C	15¼	18½	23	27½	31
D	19⅝	19⅝	19⅝	19⅝	19⅝
E	13¼	16½	21	25½	29
F	1	1	1	1	1
G	28	28	28	28	28
H	23¼	23¼	23¼	23¼	23¼
J	10¾	14	18½	23	23

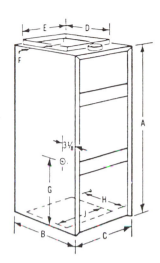

FIGURE 4-11
Furnace dimensions. (Courtesy of Southwest Manufacturing Co.)

shown in Figure 4-11. We see that furnace model CF100-BD-3 is the only one listed that will provide the airflow close to that needed for our cooling unit and the required Btu/hr capacity. The furnace chosen in our heating-only example will not deliver enough air for the air conditioning system.

4. Of the two evaporator coils listed for the condensing unit selected and shown in Figure 4-10, coil model UCC-41-23 almost exactly matches our furnace air outlet opening. Therefore, we will select this coil because very little sheet metal work will be required to connect the evaporator coil to the furnace. This sheet metal work can almost be eliminated by using an evaporator coil housing (see Figure 4-12). These housings are manufactured to accommodate the desired coil and match the furnace outlet opening. We would choose coil housing model HUH-36-48-23. Model HUH-48-61-27 would be too wide for our evaporator coil.

■ Vertical Packaged (Self-Contained) Unit Selection Procedure

The manufacturers' recommended selection procedures vary with different makes and models of self-contained equipment; however, all the procedures are basically the same. We will use the following general procedure in the example that follows.

Before beginning the selection process, the heat gain estimate must be completed and the seven items of necessary data, discussed

† Coil Housing and
Series "HL" Furnace

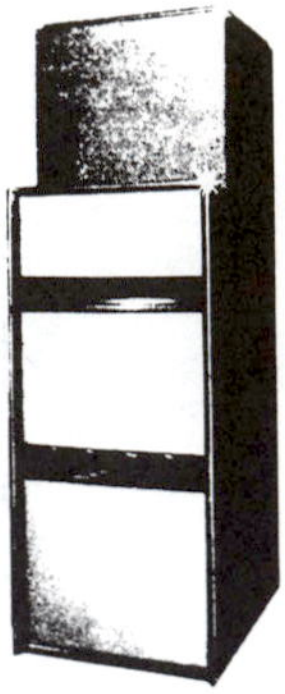

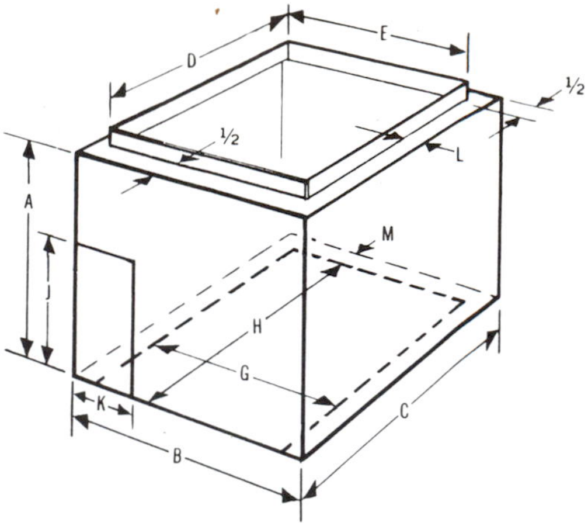

up-flow

Coil Housing MODEL NO.	Fits Coil Model No.	A	B	C	D	E	G	H	J	K	L	M
HUH-18-25-15	UCC-18-15 or 25-15	14½	15¼	21	20	14¼	13½	20⅛	11¾	4⅜	½	⅞
HUH-25-30-18	25-18 or 30-18	13¾	18½	20½	19⅜	17½	17	19¾	11¾	4⅜	½	¾
HUH-36-18	36-18	17¾	18½	20½	19⅜	17½	17	19¾	11¾	4⅜	½	¾
HUH-25-30-23	25-23 or 30-23	13¾	23	20½	19⅜	22	21½	19¾	11¾	4⅜	½	¾
HUH-36-48-23	36-23, 41-23 or 48-23	19¾	23	20½	19⅜	22	21½	19¾	11¾	4⅜	½	¾
HUH-48-61-27	48-27 or 61-27	19⅞	27½	21	19⅞	26½	26	20¼	11¾	4⅜	½	¾
HUH-48-61-31	48-27 or 61-27	19⅞	31	21	19⅞	30	26	19¾	11¾	4⅜	½	¾
HUH-90-31	UC-90-31	22¼	34⅝	25⅛	24⅛	33⅝	29½	19¾	11¾	4⅜	½	4¼

† Coil housing matches the furnace casing on sides and rear. HUH 90 overhangs 3¼ inches at rear - 1¾ on sides.

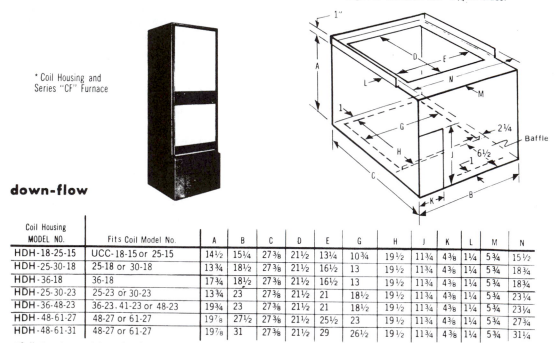

* Coil Housing and
Series "CF" Furnace

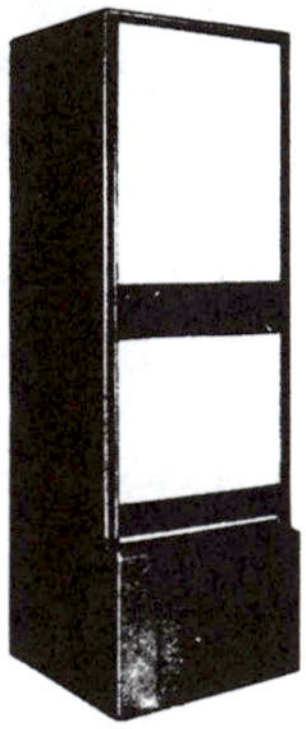

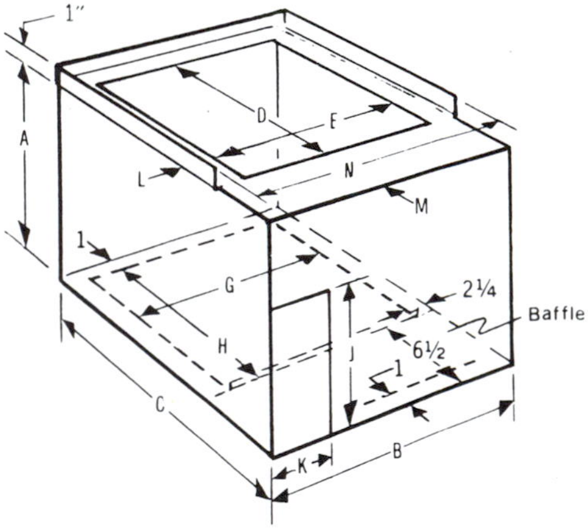

down-flow

Coil Housing MODEL NO.	Fits Coil Model No.	A	B	C	D	E	G	H	J	K	L	M	N
HDH-18-25-15	UCC-18-15 or 25-15	14½	15¼	27⅜	21½	13¼	10¾	19½	11¾	4⅜	1¼	5¾	15½
HDH-25-30-18	25-18 or 30-18	13¾	18½	27⅜	21½	16½	13	19½	11¾	4⅜	1¼	5¾	18¾
HDH-36-18	36-18	17¾	18½	27⅜	21½	16½	13	19½	11¾	4⅜	1¼	5¾	18¾
HDH-25-30-23	25-23 or 30-23	13¾	23	27⅜	21½	21	18½	19½	11¾	4⅜	1¼	5¾	23¼
HDH-36-48-23	36-23, 41-23 or 48-23	19¾	23	27⅜	21½	21	18½	19½	11¾	4⅜	1¼	5¾	23¼
HDH-48-61-27	48-27 or 61-27	19⅞	27½	27⅜	21½	25½	23	19½	11¾	4⅜	1¼	5¾	27¾
HDH-48-61-31	48-27 or 61-27	19⅞	31	27⅜	21½	29	26½	19½	11¾	4⅜	1¼	5¾	31¼

* Coil housing matches the furnace casing on sides and rear.

FIGURE 4-12
Evaporator housing dimensions. (Courtesy of Southwest Manufacturing Co.)

TABLE 4-4
Water-cooled unit data

UF Series Condensed Specifications

Model	UF036W	UF060W	UF090W	UF120W	UF180W	UF240W	UF300W	UF360W
Dimensions (in.): Height	78	78	78	78	78	79 1/2	79 1/2	79 1/2
Width	30	42	54	65	89	76 1/4	96 1/2	96 1/2
Depth	20	20	20	28	28	38 1/2	38 1/2	38 1/2
Ship Wt. (lbs.)	370	510	672	1020	1267	1445	1865	2025
Filters— †Disposable ‡Permanent, cleanable	†(1)20×25×1	†(1)20×25×1 †(1)16×25×1	†(3)16×25×1	‡(3)19×27×1	‡(8)15×20×1	‡(4)20×25×1 (2)20×20×1	‡(6)25×30×1	‡(6)25×30×1
Performance: Cooling, std rating (Btuh)	36,000	60,000	86,000	120,000	176,000*	240,000*	300,000*	360,000*
Power input (watts)	3,950	6,700	9,700	13,100				

*Units above 120,000 Btuh not ARI rated

UR Series Condensed Specifications

Model	UR030	UR036	UR048	UR060
Dimensions (in.): Height	26 11/32	26 11/32	30	30
Width	42 15/16	42 15/16	52 3/8	52 3/8
Depth	45 3/16	45 3/16	52 3/8	52 3/8
Ship Wt. (lbs.)	320	340	450	510
Filters (recommended), dimensions (in.) not included in unit	(2)16×20×1	(2)16×20×1	(2)20×20×1	(2)20×20×1
Performance: Cooling, standard rating (Btuh)	30,000	38,000	46,000	61,000
Power input (watts)	4,300	5,100	5,400	8,300

Electrical data:	UR030		UR048		
Phase/Cycle	1/60		1/60	3/60	3/60
Rated voltage	208-230		230	230	460
	UR036		UR060		
Phase/Cycle	1/60	3/60	1/60	3/60	3/60
Rated voltage	230	208-230	230	208-240	460

under the heading "Data Required for Cooling Unit Selection," must be gathered. Once these tasks are completed, the packaged unit may be selected.

Selection.

1. The first step is to make a tenative selection based on the heat gain calculation. This selection is made from the manufacturer's nominal unit capacity (see Table 4-4).
2. Compare the rated cfm air delivery for the unit selected, when provided in the table, to the air delivery requirement for the building. If the rated cfm is within 20% of the required cfm, the selection should be made on the basis of the required cfm. If there is a difference of more than 20% in the cfm, a special builtup system may be considered.
3. Calculate the entering dry bulb and wet bulb temperatures, or obtain them from a table (see Table 4-3). Regardless of which method is used, the indoor and outdoor design temperatures and the ventilation requirements must be known.

4. When water-cooled condensers are used, the condensing temperature should be determined. A tentative condensing temperature of approximately 30°F higher than the entering water temperature is normally taken.

5. Determine the total unit capacity and the sensible heat capacity of the unit at the conditions given for the unit selected. This information is usually listed in the manufacturers' equipment tables.

   a. When selecting water-cooled units, the evaporator entering air dry bulb and wet bulb temperatures and the calculated condensing temperature are necessary.

   b. When considering air-cooled units, the outdoor design dry bulb temperature, and the dry bulb and wet bulb temperatures of the air entering the evaporator are necessary. Use the smallest condenser recommended at this point of the selection process. When the capacity of the unit selected falls between the ratings listed in the table, it will be necessary to interpolate to obtain the exact capacities of the unit. The unit capacities listed in these tables are for standard air delivery only. When quantities other than standard are to be used, correction factors must be applied to the listed capacities. These correction factors are listed in tables (see Table 4-5).

6. The total capacity of the unit selected should be equal to or greater than the total cooling load unless the design temperatures can be compromised. When additional capacity is required, a new selection must be made to handle the load.

   a. When considering a water-cooled condenser, the condensing temperature should be lowered 5 to 10°F and the new unit capacities should be considered before selecting a larger unit. However, if the load remains larger than the unit capacity, the next larger unit must be used.

TABLE 4-5
Capacity Correction Factors*

CFM COMPARED TO RATED QUANTITY	−40%	−30%	−20%	−10%	STD.	+10%	+20%
Cooling Capacity Multiplier	88	93	96	98	1.00	1.02	1.03
Sensible Capacity Multiplier	80	85	90	95	1.00	1.05	1.10
Heating Capaciity Multiplier	78	83	89	94	1.00	1.06	1.12

*To be applied to cooling and heating capacities

b. When considering an air-cooled condenser and the unit capacity will not meet the load requirements, a larger-capacity condenser should be selected. Should the unit capacity still be insufficient, the next-larger-size unit must be selected.

7. After establishing the actual unit capacity, compare the sensible heat capacity of the unit to the heat gain load of the building. If the unit sensible heat capacity is not equal to or greater than the cooling load, a larger unit must be selected.

8. When considering water-cooled condensing units, determine the gpm water flow required and the water pressure drop through the condenser. Use the final condensing temperature and the total cooling load in tons of refrigeration to determine the gpm requirement from a condenser water requirement table, when provided for the unit being considered. After the condenser gpm requirement is determined, the pressure drop can be read from the table prepared by the manufacturer, if this information is given.

9. Determine the fan speed and the fan motor horsepower required to deliver the required volume of air against the calculated external static pressure caused by the air distribution system. The air volume is read directly from the manufacturer's charts and performance tables. Should the required fan motor horsepower exceed the recommended limits, a larger fan motor must be used.

Vertical self-contained units are usually available only in fairly large capacity increments. Many situations are encountered when the actual heat gain load falls between two of these capacities. For example, the calculated heat gain load indicates a 12.5-ton load. The available units are either 10 or 15 tons of capacity. With the possibility of a future expansion, the building owner may authorize a single 15-ton unit. However, if he is economy minded, he may decide to tolerate a few days which are warmer than normal and authorize the smaller 10-ton unit. The contractor can, however, offer a 7.5-ton unit and a 5-ton unit to match the load exactly. In this case the customer must be willing to pay the higher cost of a two-unit installation.

It is always wise, whatever the case may be, to compare the actual sensible and latent heat capacities of the unit with the calculated heat gain of the building. It is often most desirable, for comfort reasons, that the unit have the capacity to handle the complete sensible heat load.

When the sensible heat percentage (the ratio of sensible heat gain to latent heat gain) is normally high or low, a more detailed check should be made. This check can be made by consulting the equipment manufacturer's capacity tables.

■ Vertical Packaged (Self-Contained) Equipment Selection Procedure

Let us select a vertical self-contained water-cooled condensing unit for use in an office building in Dallas, Texas. The condensing water is to be supplied from a cooling tower. The unit will be used with a free discharge air plenum. No specified air delivery requirements. The necessary data are:

1. The outdoor design temperatures are 100°F DB and 78°F WB.

2. The indoor design temperatures are 78°F DB and 65°F WB.

3. The ventilation air required is 300 cfm.

4. The total cooling load is 7 tons (84,000 Btu/hr).

5. The condensing medium is water at 85°F.

6. The air delivery required is 400 cfm/ton or 2800 cfm.

7. External static pressure is none.

Selection.

1. From the manufacturer's rating chart, select unit model UF090W (see Table 4-4).

2. When supplied, determine the rated cfm delivery for the model chosen. This will normally be 400 cfm/ton. In our case the amount of air delivered would be 2866 cfm.

3. Calculate the DB and WB temperatures of the air entering the evaporator:

   a. $\dfrac{\text{Ventilation requirement}}{\text{Standard cfm delivery}} = \dfrac{300}{2866} = 10.4\%$ or 10% outside air

   b. Entering air DB (°F) = $(0.10 \times 100°F) + (0.90 \times 78°F) =$ 82.2°F DB
   (°C) = $(0.10 \times 37.78°C) + (0.90 \times 25.56°C) =$ 26.8°C DB

   c. The WB temperature at 80.2°F DB with the given mixture is found on the psychrometric chart to be 66.4°F WB.

4. Calculate the condensing temperature: 85°F + 30 = 115°F condensing temperature.

5. Because there is no specified latent heat load, the sensible heat percentage is not considered in our example.

6. Determine the gpm water flow required. When the manufacturer's table does not include this information, it is customary to figure 3 gpm per ton of refrigeration. Thus $7.16 \times 3 = 21.49$ gpm.

7. Determine the air delivery in cfm of the unit. When this information is not included in the manufacturer's table, it is customary to figure 400 cfm per ton of refrigeration. Thus $7.16 \times 400 = 2864$ cfm. The air can now be properly distributed to the needed areas.

■ Horizontal Packaged (Self-Contained) Unit Selection Procedure

When considering horizontal self-contained units for commercial installations, essentially the same procedure is used as when selecting vertical air-cooled self-contained units. However, for residential installations, the ventilation requirements will not need to be considered.

Before beginning the selection process, the heat gain estimate must be completed and the seven items of necessary data, discussed under the heading "Data Required for Cooling Unit Selection," must be gathered. Once these tasks are completed, the packaged unit may be selected.

Selection.

1. From the manufacturer's capacity table, select a unit based on the total calculated heat gain calculation.

2. Determine the standard cfm rating of the unit selected, if included, and compare it to the cfm required for the building.

3. Determine the WB and DB temperatures of the air entering the evaporator. Use the formula

$$\text{Entering air DB} = (\% \text{ outside air} \times \text{outdoor DB}) + (\% \text{ inside air} \times \text{indoor DB})$$

This was done as step 3 in the section headed "Vertical Packaged (Self-Contained) Unit Selection Procedure."

4. Determine the total unit capacity and sensible heat capacity at the various conditions for the unit selected. This information is obtained from the manufacturer's capacity table, when provided, using the design outdoor DB and the return air DB and WB temperatures.

5. When the total unit capacity is not sufficiently close to the calculated heat gain, another unit must be selected.

The capacity of air-cooled units cannot be adjusted as readily as water-cooled units. Nor do air-cooled packaged units offer a choice of evaporator or condenser coil for varying the system capacity. This problem is alleviated, however, because air-cooled packaged units are readily available in a larger number of capacity ranges. When the calculated heat gain cannot be met exactly, the alternate choice of a larger or smaller unit, or a multiple-unit type of installation will depend on the owner's preference. The considerations for making this decision are basically the same as those listed for vertical self-contained units.

SUMMARY

The proper sizing and selection of air conditioning equipment is just as important as any other step in the estimating process.

The proper sizing and selection of air conditioning equipment can be accomplished only after an accurate heat loss and heat gain analysis is completed.

The selection of equipment is a problem of choosing the unit that is the closest to satisfying the demands of capacity, performance, and cost.

The information that must be determined before attempting to select a warm air furnace includes such items as (1) the inside and outside design temperatures, (2) the total heat loss in Btu/hr, (3) the type of heating equipment to be used, (4) the required cfm, and (5) the external static pressure caused by the ductwork.

When considering the design temperatures, only the dry bulb temperatures are required.

The total heat loss figure is determined by completing a heat loss form.

The type of furnace selected will depend on several application variations, such as the type of energy available, or desired; and whether an upflow, downflow, or horizontal airflow is desired.

The direction of airflow and the type of energy chosen must be known before the furnace manufacturers' tables can be used.

To distribute the heated air properly in the conditioned space will require that the cfm be properly calculated before the duct system is designed.

The external static pressure at the required cfm is a vital part of the complete duct system design calculations. It should, therefore, be included in the duct system specifications.

When the total heat loss has been calculated, there are only two steps left in selecting a furnace for a heating-only system: (1) selecting the furnace from the manufacturer's tables, and (2) determining the necessary cfm adjustment for heating.

The balancing of system components, such as the condensing unit and the evaporator, and the air delivery and the duct system, is important if comfort and maximum efficiency are to be maintained.

The selection of cooling equipment, whether it be a remote, split system or a packaged unit, has been made relatively simple because the equipment manufacturers have already determined what components will produce a given unit capacity.

The following information must be known before the proper selection of cooling equipment for residential or light commercial installations can be completed: (1) the summer design outdoor air dry bulb (DB) and wet bulb (WB) temperatures, (2) the conditioned space design DB and WB temperatures, (3) the required ventilation air in cfm, (4) the total cooling load in Btu/hr, (5) the cfm air delivery required, (6) the condensing medium (air or water) and the temperature of the medium entering the condenser, and (7) the duct external static pressure.

The design outdoor air temperatures used should be those commonly used in the geographical area where the building is located.

The conditioned space design temperatures are also necessary for two reasons: (1) the load calculations, and (2) the evaporator entering-air-temperature calculation.

The required cfm of ventilation air must be known before attempting to determine the temperature of the air entering the evaporator or attempting to determine the sensible heat capacity of the air conditioning system.

The total cooling load (heat gain) is the basis for the complete air conditioning unit selection.

When the long-form heat gain calculation form is used, the cfm required can be calculated from the total sensible heat load and the rise in the DB temperature of the air supply. On some load calculation forms no cfm requirement is specified. In such cases, the unit is chosen on the Btu/hr capacity alone.

When water-cooled condensers are used with well water or city water, the entering water temperature must be determined to calculate properly the gpm of water that must flow through the condenser to achieve the design condensing temperatures.

The capacity of an air-cooled condenser is determined on the difference between the condensing temperatures and the design outdoor air DB temperatures.

When an air conditioning unit is to be connected to a duct system, the external static pressure must be known to determine the fan speed and the fan motor horsepower required.

After the heat gain calculations have been completed, the split-system equipment can then be chosen according to the appropriate following application and installation considerations: (1) the temperature of the air entering the evaporator and condenser are determined, (2) a split system with sufficient capacity is selected from the manufacturers' tables, and (3) the air delivery by the unit against the calculated external static pressure must be determined.

REVIEW QUESTIONS

1. How is the size of air conditioning equipment determined?

2. What will be the result of an air conditioning unit that is sized too small?

3. What will be the result of installing an air conditioning unit that is too large?

4. List the data required for heating unit selection.

5. When considering the inside and outside design temperatures for heating, what temperatures are required? (WB or DB)

6. How is the total heat loss figure determined?

7. What is the most commonly used warm air furnace used in residential installations?

8. What type of forced warm air furnace has the blower on top blowing downward?

9. In what types of installations are horizontal furnaces ideal?

10. Name two sources for furnace vent diameter data.

11. List the data required for cooling unit selection.

12. List two reasons it is necessary to know the design outdoor air conditions.

13. Where can the design outdoor air temperatures be found?

14. List two reasons why the conditioned space design DB and WB temperatures are necessary.

15. What is the basis for selecting a complete air conditioning system?

16. What is the assumed cfm air delivery of an air conditioning unit?

17. List the application and installation considerations for split-system equipment selection.

5 Equipment Location

In the first four chapters we have discussed psychrometrics, heat loss, heat gain, and equipment sizing and selection. In this chapter we discuss some principal considerations regarding the location and installation of the equipment.

INTRODUCTION

A well-planned location for the air conditioning and heating unit is one of the most important steps in a good installation. The following are some of the considerations that are involved in planning the location of the various units.

FURNACES

There are three types of furnaces when classified by their source of energy: (1) gas, (2) electric, and (3) oil. The gas furnaces are further broken down according to the direction of air flow through them. They are: upflow, downflow (counterflow), and horizontal (see Figure 5-1).

HI-LO   COUNTER-FLO   HORIZONTAL

FIGURE 5-1
Upflow, counterflow, and horizontal gas fur-
naces. (Courtesy of Southwest Manufacturing
Co.)

Electric furnaces are more flexible in installation and thus are not subdivided by the direction of airflow through them (see Figure 5-2).

Oil furnaces are generally classified further by the direction of air-flow through them, like the gas furnaces listed above.

■ Gas Furnaces (General)

All gas furnaces require some of the same considerations with regard to location and installation. Thus we will discuss at this time such items as gas pipe connections, venting requirements, return air location, supply air location, minimum clearances, noise, accessibility, length of duct runs, and thermostat location.

Upflow   Horizontal   Conterflow

FIGURE 5-2
Upflow, horizontal, and counterflow electric fur-
nace positions. (Courtesy of Southwest Manufac-
turing Co.)

Gas Pipe Connections. The availability of the gas pipe to the furnace is important from a labor viewpoint. When the gas pipe is already in the equipment room, a certain amount of labor and materials will be saved. However, should the gas pipe be located on the other side of the building, the cost of labor and materials will be greatly increased, thus increasing the cost of the installation. The gas supply to the building must have the capacity to handle the extra load that the new equipment will place on the supply piping. Be sure to consider any restrictions or requirements that might be required because of local or national building codes.

Venting Requirements. When considering the venting requirements for gas-fired furnaces, any and all local and national codes and ordinances must be followed. An unsafe furnace should not be installed or operated because of the possible hazard to life and property. The furnace should be located so that the vent pipe will provide as straight a path as possible for the flue gases to escape to the atmosphere. Never decrease the size of vent piping from that recommended by the manufacturer. Vent pipe is very expensive; therefore, a short, straight run is not only easier but also more economical to install (see Figure 5-3).

When the furnace is to be installed in a closet or tightly closed room, openings must be provided to ensure that adequate combustion air is available to the unit. Be sure that the local and national codes are followed when locating these openings and their size. It is more desirable from a safety viewpoint and for more economical operation if

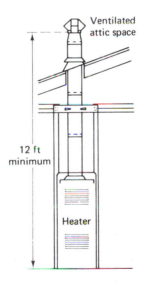

FIGURE 5-3
Minimum vent height. (Courtesy of Selkirk Metalbestos, Division of Household International)

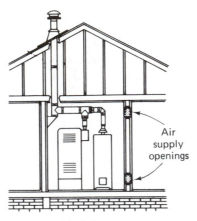

All air from
inside building

$$\frac{\text{Free area of}}{\text{each grill}} = \frac{\text{Total input}}{1000}$$

(Use 2 grills facing into
large interior room)

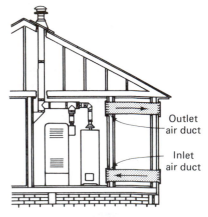

All air from outdoors

$$\frac{\text{Free area of}}{\text{each duct}} = \frac{\text{Total input}}{2000}$$

$$\frac{\text{Free area of}}{\text{each grill}} = \frac{\text{Total input}}{4000}$$

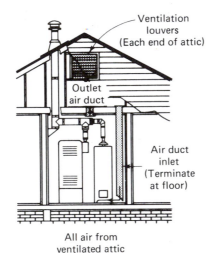

All air from
ventilated attic

$$\frac{\text{Free area of each duct or grill}}{} = \frac{\text{Total input}}{4000}$$

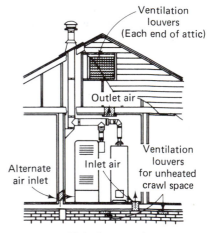

Air in from crawl
space, out into attic

$$\frac{\text{Free area of}}{\text{each grill}} = \frac{\text{Total input}}{4000}$$

FIGURE 5-4
Suggested methods of providing air supply.
(Courtesy of Selkirk Metalbestos, Division of
Household International)

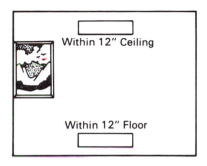

FIGURE 5-5
Combustion air openings in equipment room wall.

the combustion air can be taken from outside the living space (see Figure 5-4).

However, if it is not feasible to duct the combustion air from outside, it may be taken from inside the living space. In either case, approximately half the air should be introduced near the ceiling while the other half is introduced near the floor (see Figure 5-5). Adequate covering must be placed on the openings to prevent insects from entering through the duct or opening. Be sure to provide extra dimensions to the combustion air intake because of the added restriction caused by the grill or covering.

Return Air Location. There are several methods used for directing the return air to the furnace. The most popular are: a central return (see Figure 5-6), a return in each room (see Figure 5-7), and a combination of the two. It is obvious that a return air duct to each room would require more material and labor, thus increasing the cost of the installation. However, it is sometimes more desirable than the central return because of the added comfort and reduced noise.

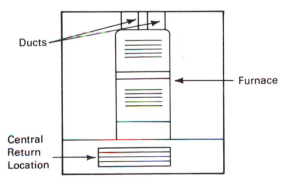

FIGURE 5-6
Central return location.

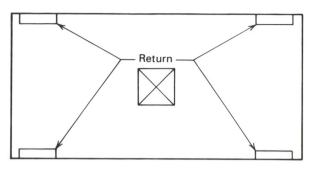

FIGURE 5-7
A return outlet in each room.

Under no circumstances should the equipment closet be used as the return air plenum. The return air duct and plenum should completely isolate the return air from the combustion air in the closet. This precaution is to prevent the products of combustion from entering the living space.

Supply Air Location. The supply air plenum is located on the air outlet of the furnace. The air distribution system is connected to the plenum and directs the air to the desired points in the living space. There are several types of duct systems and the one that is most desirable for the particular installation should be used (see Figure 5-8).

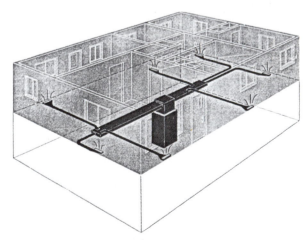

(a) Extended plenum system

FIGURE 5-8
Various types of duct systems.

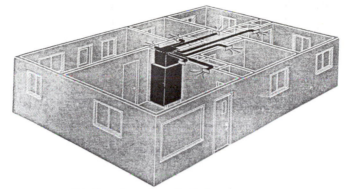

(b) Overhead radial duct system

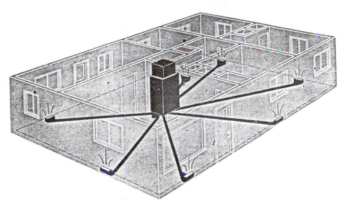

(c) Radial perimeter system

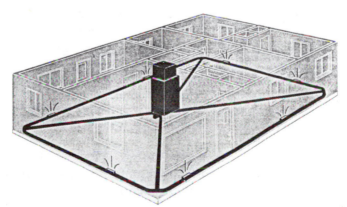

(d) Loop perimeter system

FIGURE 5-8 (Continued)

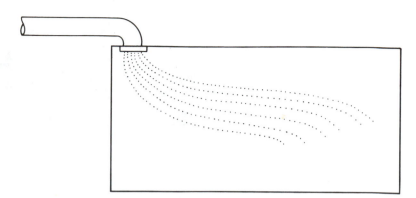

FIGURE 5-9
Supply air outlet location.

Naturally, when more material and labor are required, the price of the installation will increase accordingly.

The supply openings should be located so that the air will be evenly distributed throughout the room, or where the air will be directed to the point of greatest heat loss or gain. Generally, the air will be directed toward an outside wall or window (see Figure 5-9). The location of the supply opening will determine, to a great extent, the length of duct required.

Minimum Clearances. Under no circumstances should combustible material be located within the minimum clearances given in the manufacturer's installation literature. These clearances are set by the American Gas Association for each model of furnace and they must be adhered to. The furnace dimensions should be correlated with the closet or space dimensions to ensure that sufficient room is available. If not, another location must be found or the room size increased to accommodate the furnace safely.

Noise. Because of air noise and equipment vibration, the unit should not, if at all possible, be located near bedrooms or the family room. However, this will usually increase the installation costs because of the added return air duct required. When such a location cannot be avoided, acoustical materials can be placed in the return air plenum and the air handling section of the equipment to reduce the air noise. Vibration can be reduced, or eliminated, by placing the unit on a solid foundation and using vibration-elimination materials.

Accessibility. When considering possible locations for the equipment, sufficient room must be allowed for servicing the unit. A minimum of 24 in. in front of the unit is necessary for changing filters, lubricating bearings, lighting the pilot, and so on. In some installations more than 24 in. is required for servicing procedures.

Length of Duct Runs.   The indoor unit should be located so that the length of the duct runs to each room will be as short as is possible. The number of turns and elbows must also be kept at a minimum. These precautions help to reduce heat loss and heat gain through the air distribution system, and allow simpler design and installation. The shorter duct runs also help to reduce the amount of resistance to airflow, allowing the air delivery to be balanced much more easily.

Thermostat Location.   Satisfactory operation of an air conditioning system depends greatly on careful placement of the thermostat. Following are points that must be considered when locating a thermostat:

1. It must be in a room that is conditioned by the unit it is controlling.

2. It must be exposed to normal free air circulation. Do not locate the thermostat where it will be affected by lamps, appliances, fireplaces, sunlight, or drafts from the opening of doors or outside windows.

3. It must be on an inside wall about 4 to 5 ft from the floor.

4. It must not be behind furniture, drapes, or other objects that would affect the normal flow of free air.

5. It must not be located where it can be affected by vibrations, such as on a wall of an equipment room, or on the wall in which the door is located.

■ Gas Furnaces (Upflow)

Upflow furnaces are best suited for closet, garage, or basement installations (see Figure 5-10). When an air conditioning unit is to be included, drainage facilities must be available. Otherwise, disposal of the condensate water will be a problem that could be expensive. The floor or foundation where the unit will set should be level, dry, solid, and strong enough to support the weight of the unit. If supply and return air is to be ducted to the unit, a central location should be chosen to ensure efficient and economical air distribution. Check the manufacturer's installation information to determine clearances and other pertinent data.

■ Gas Furnaces (Downflow)

These furnaces are best suited for closet, garage, or in second-floor attics (see Figure 5-11). Basically the same considerations apply to

FIGURE 5-10
Upflow furnace. (Courtesy of Southwest Manufac-
turing Co.)

downflow furnaces as to upflow furnaces. The main exception is that there be sufficient room beneath the furnace and building to install the supply air duct and plenum. These types of systems usually distribute the air from the floor upward. However, when installed in a second-story attic the air may be delivered to the first floor from the ceiling. The ducts are generally placed either beneath the floor, in the slab foundation, or between the second-story floor and the first-floor ceiling. Adequate condensate drain facilities should be available or a more expensive drain system will be needed when used with a cooling system.

■ Gas Furnaces (Horizontal)

Horizontal furnaces are best suited for crawl space or attic installations (see Figure 5-12). When they are located in the crawl space, the air is generally supplied to the space through the floor. In the attic, the air is supplied through the ceiling. Installation of the vent pipe is usually more difficult and expensive with crawl space installation than with other types. Also for this type of furnace, there must be adequate access for servicing. Otherwise, poor service with reduced efficiency and safety will result. These units are generally suspended from the floor joists or the rafters. Horizontal furnaces are usually more diffi-

FIGURE 5-11
Counterflow furnace. (Courtesy of Southwest
Manufacturing Co.)

cult to install and service than the other types because of the equipment location. When used in conjunction with a cooling system, an auxiliary condensate drain pan and line are required to help prevent water damage to the ceiling should the drain become clogged on attic installations. However, the local codes and ordinances will dictate these requirements.

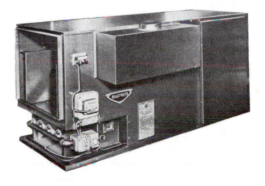

FIGURE 5-12
Horizontal furnace. (Courtesy of Southwest Man-
ufacturing Co.)

FIGURE 5-13
Electric furnace. (Courtesy of Southwest Manu-
facturing Co.)

◼ Electric Furnaces

There is much more flexibility when locating electric furnaces than with gas furnaces, because there is no need for vent or gas pipe installation (see Figure 5-13). However, it is more economical to locate an electric furnace as close to the electric service panel as possible. The Btu/hr rating of electric furnaces can be changed by adding or subtracting the number of heating elements. They may be installed in virtually any position because there are no vent piping or burner requirements to consider.

The same general requirements as to return air location, supply air location, minimum clearances, noise, accessibility, length of duct runs, and thermostat location for gas furnaces apply to electric furnaces. When these units are used in combination with cooling systems, adequate drainage should be available or provisions made to dispose of the condensate water.

COOLING EQUIPMENT

Basically, there are two types of cooling systems: split-system and self-contained (packaged) units. The split-system evaporator and condensing unit are in different locations. The self-contained unit has the evaporator and condensing unit in one cabinet. These systems may be

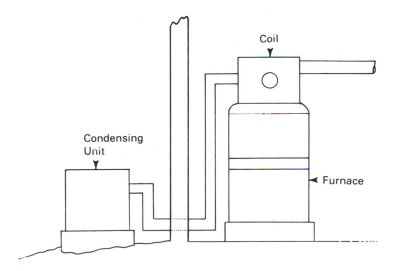

FIGURE 5-14
Split-system installation.

operated as cooling-only units or as heating and cooling units when used in combination with a heating apparatus.

■ Split-System Units

These types of systems usually have the evaporator installed inside the conditioned space and the condensing unit placed outside at a remote location (see Figure 5-14). The evaporator may be located inside a cabinet with a blower or mounted on the air outlet end of a furnace (see Figure 5-15).

The following considerations should be taken into account when choosing the location for split-system units: support, access, air supply, comfort, length of refrigerant lines, and length of electric lines. These are in addition to those listed for the various types of furnaces discussed earlier in the chapter.

Support.   When the condensing unit is placed on the ground, a 4-in. concrete slab should be used to support the unit (see Figure 5-16). When the unit is to be mounted on the roof or other parts of the structure, the construction must be checked to make certain that it has sufficient strength to support the weight of the unit.

Access.   Two or three feet of clearance should be provided on the side where the access panel is located. This clearance is necessary for

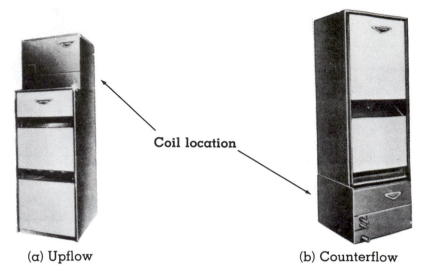

Coil location

(a) Upflow   (b) Counterflow

FIGURE 5-15
Upflow and counterflow coil locations. (Courtesy
of Southwest Manufacturing Co.)

routine servicing operations. There should be sufficient clearance around the unit to allow for airflow to the condenser. Check the manufacturer's instructions for the clearance required.

Air Supply.   When the desired location offers a restricted air flow, such as in a garage or carport duct, work may be required to allow a sufficient amount of air to flow across the condenser coil and out of the fan discharge. When a horizontal discharge unit is installed, it may be desirable to install an air deflector to direct the hot air away

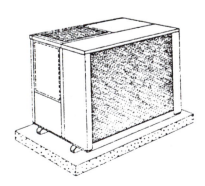

FIGURE 5-16
Concrete slab on ground.

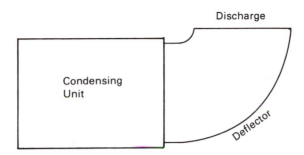

FIGURE 5-17
Horizontal discharge condenser with an air deflector.

from grass, shrubs, flowers, or a neighbor's property (see Figure 5-17).

When vertical discharge condensing units are used, the unit must be located so that the discharge air will not be recirculated back into the unit. This usually requires that the unit be located far enough from the house so that the air will miss the roof eaves (see Figure 5-18). In such installations the refrigerant and electrical lines should be buried underground to prevent damage to them.

Comfort Considerations. The condensing unit should not be located close to bedrooms or close to a patio, where noise and the discharge air would be objectionable. The unit should not interfere with a neighbor's property or personal convenience. It is usually preferred to locate the condensing unit by a bathroom.

Length of Refrigerant Lines. The condensing unit should be located as close as possible to the evaporator to reduce the length of the refrigerant lines. Long refrigerant lines are more susceptible to refrigerant

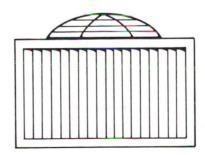

FIGURE 5-18
Vertical discharge condenser.

leaks than are shorter ones. Also, the cost of the refrigerant lines will be less. With shorter lines the initial refrigerant charge will be smaller, and if a leak should occur, a smaller quantity of refrigerant will be lost. An increased restriction to the flow of refrigerant will be experienced with longer refrigerant lines. This restriction must be kept at a minimum.

Length of Electric Lines. The location of the electrical service panel in relation to the unit should be considered. The longer the electric lines, the more costly the installation. However, a long electric line should not outweigh long refrigerant lines. Always check the manufacturer's specifications for the size of wiring for each distance from the electric supply panel.

■ Self-Contained (Packaged) Units

Self-contained air conditioning equipment is manufactured in basically two types: upflow and horizontal units. The type of design chosen will depend greatly on the installation location. When air-cooled units are located indoors or in through-the-wall installations, a sufficient amount of clearance must always be allowed in front of the condenser air intake to allow plenty of air to pass over the condenser coil. A grill or some type of screen should be installed on the air intake to prevent the entrance of debris and foreign matter into the unit.

When water-cooled units are located indoors, there should be an accessible point in the structure through which the water lines can be installed. Otherwise, a much longer path must be taken, which will increase the initial installation costs and place an added restriction on water flow through the pipes.

Upflow Self-Contained Units. The location chosen for upflow self-contained units must have adequate electrical and water supplies available. It is desirable to have drainage facilities close to the unit to prevent the extra expense of installing a drain pump. The floor or the foundation where the unit will be located should be strong enough to support the weight of the unit. It should be level, dry, and solid. When supply and/or return ducts are to be installed, the unit should be installed in a central location to ensure that efficient and economical air distribution can be obtained. There should be adequate clearance around the unit so that routine maintenance can be provided. The manufacturer's installation literature usually will provide the weight, necessary clearances, and other pertinent data about the unit.

Horizontal Self-Contained Units. When horizontal self-contained units are under consideration, many of the location considerations listed for upflow packaged units, split-system units, and furnaces will apply. These types of units are generally used on mobile homes, in crawl spaces, and for through-the-wall installations. Remember that a central location is best for air distribution purposes. In addition, electrical power and drain facilities, together with the strength of the structural support, should be considered. Sufficient clearance must be maintained around the unit for servicing procedures. Avoid locating the equipment close to sleeping areas.

Example Location. In locating the unit for our example, the architect designed the indoor unit location at the door to bath 1 in the furnace closet (see Figure 5-19). The air conditioning unit design engineer can place the condensing unit in the best location, keeping in mind the facts discussed earlier in this chapter. If this is new construction, there is not much of a problem because the refrigerant lines and electric wires can be installed without any problem during the construction of the building.

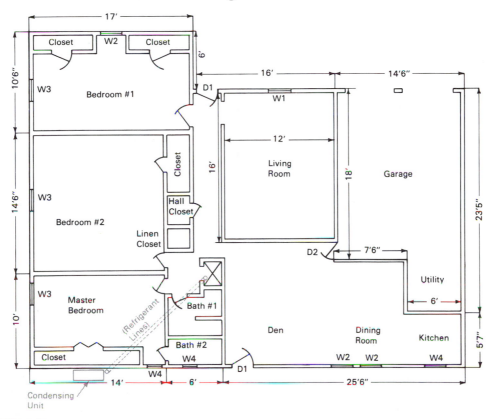

FIGURE 5-19
Condensing unit and indoor unit location.

However, if the unit is being installed in an existing home, consideration must be given to such things as type of floors, amount of room in the attic, ease of installation of ducts and refrigerant lines, easy access to a condensate drain, and so on.

In our example let us assume that the unit is to be installed as the building is under construction. In this case, the refrigerant lines would probably be installed in a chase beneath the slab floor to shorten the lines and make them easier to install. In keeping with the aforementioned installation practices, the best location for the condensing unit is on the east wall of the master bedroom about midway of the closet. The noise level would be at a minimum and the refrigerant lines would not be unnecessarily long. We would not want to put the unit outside bath 2 because of the window to the bath and master bedroom and the door to the den. All of these would allow more noise to enter the house than could enter through the outside wall and the inside closet wall. Also, the discharge air from the condenser might be objectionable if it blew across these items when opened.

SUMMARY

There are three types of furnaces as classified by source of energy: (1) gas, (2) electric, and (3) oil.

Gas and oil furnaces are further classified according to the direction of airflow through them: upflow, downflow (counterflow), and horizontal.

Electric furnaces are not subdivided by the direction of airflow through them.

Items such as gas pipe connections, venting requirements, return air location, supply air location, minimum clearances, noise, accessibility, length of duct runs, and thermostat location should be considered when locating a gas furnace.

When considering the venting requirements for a gas-fired furnace, any and all local and national codes and ordinances must be followed. An unsafe furnace should not be operated because of the hazard to life and property.

When a gas furnace is to be installed in a closet or a tightly closed room, openings must be provided to ensure that adequate combustion air is available to the unit.

The central return is the most popular method of directing the return air to the furnace because of the added cost of return air ducts.

Under no circumstances should the equipment closet be used as the return air plenum.

The supply air plenum is connected to the air outlet of the furnace.

The air distribution system is connected to the plenum and directs the air to the desired points in the living space.

The supply air openings should be located so that the air will be evenly distributed throughout the room, or where the air will be directed to the point of greatest heat loss or gain.

Under no circumstances should combustible material be located within the minimum clearances given in the manufacturer's literature.

Because of air noise and equipment vibration, the unit should not be located near bedrooms or the family room if at all possible.

When considering possible locations for the equipment, sufficient room must be allowed for servicing the unit.

The indoor unit should be located so that the length of the duct runs to each room will be as short as possible. The number of turns and elbows must also be kept to a minimum. These precautions help to reduce heat loss and heat gain through the air distribution system.

The satisfactory operation of an air conditioning system depends greatly on careful placement of the thermostat.

Upflow furnaces are best suited for closet, garage, or basement installations.

Downflow furnaces are best suited for closet, garage, or in second-floor attics.

Horizontal flow furnaces are best suited for crawl space or attic installations.

There is much more flexibility in locating electric furnaces than with gas furnaces, because there is no need for vent pipe or gas pipe installation.

The Btu/hr rating of an electric furnace can be changed by adding or subtacting the number of heating elements.

Basically, there are two types of cooling systems: split-system and self-contained (packaged) units. The split-system evaporator and condensing unit are in different locations. The self-contained unit has the evaporator and condensing unit in one cabinet.

The following considerations should be taken into account when choosing the location for split-system units: support, access, air supply, comfort considerations, length of refrigerant lines, and length of electric lines.

Horizontal self-contained units are generally used on mobile homes, in crawl spaces, and for through-the-wall installations.

REVIEW QUESTIONS

1. Why are electric furnaces more flexible than gas furnaces?

2. What should be followed when considering the venting requirements of a gas furnace?

3. Why should an unsafe furnace not be installed or operated?

4. Is it good practice to decrease the size of vent piping from that recommended by the manufacturer?

5. Where should the combustion air to a gas furnace be introduced into the closet?

6. Should the equipment closet be used as the return air plenum?

7. To what area should the supply air openings direct the air?

8. What organization sets the minimum clearances on a gas furnace?

9. Is it desirable to locate a unit close to bedrooms or the family room?

10. How can vibration be reduced or eliminated from a unit?

11. In what room should a thermostat be placed?

12. How far from the floor should a thermostat be mounted?

13. For what type of installations are upflow furnaces best suited?

14. What type of furnace is best suited for crawl spaces or attic installations?

15. Name two basic types of cooling systems.

6 Refrigerant Lines

When installing units with remote condensers or other components, it will be necessary to connect these components with field-engineered and field-installed refrigerant piping. When a standard catalog combination of units is being considered, the piping size calculations will have already been made by the equipment manufacturer's engineers. The pipe diameter and maximum lengths will be listed in the installation literature. When a noncatalog combination such as a 10-ton condensing unit with two 5-ton evaporators is being considered, the line sizes should be calculated by the methods that are discussed in the remainder of this chapter.

INTRODUCTION

Refrigerant piping design procedures are a series of compromises, at best. A good piping system will have a maximum capacity, be economical, provide proper oil return, provide minimum power consumption, require a minimum amount of refrigerant, have a low noise level, provide proper refrigerant control, and allow perfect flexibility in system performance from 0 to 100% of unit capacity without lubrication problems. Obviously, it is impossible to obtain all these goals because some of them are in direct conflict with others. To make an intelligent decision as to what type of compromise is most desirable, it is essential that the piping designer have a thorough understanding of the basic effects on the system performance of the piping design in the different points of the refrigerant system.

PRESSURE DROP

Pressure drop, in general, tends to decrease system capacity and increase the amount of electric power required by the compressor. Therefore, excessive pressure drops should be avoided. The amount of the pressure drop allowable will vary depending on the particular segment of the system that is involved. Each part of the system must be considered separately. There are probably more charts and tables available which cover refrigeration line pressure drop and capacities at a given pressure, temperature, and pressure drop than any other single subject in the field of refrigeration and air conditioning.

The piping designer must realize that there are several factors which govern the sizing of refrigerant lines; pressure drop is not the only criterion to be used in designing a system. It is often required that refrigerant velocity, rather than pressure drop, be the determining factor in system design. Also, the critical nature of oil return can produce many system difficulties. A reasonable pressure drop is far more preferable than oversized lines which may hold refrigerant far in excess of that required by the system. An overcharge of refrigerant can result in serious problems of liquid refrigerant control, and the flywheel effect of large quantities of liquid refrigerant in the low-pressure side of the system can result in erratic operation of the refrigerant flow control device.

The size of the refrigerant line connection on a service valve that is supplied with a compressor, or the size of the connection on an evaporator, condenser, or some other system accessory does not determine the correct size of the refrigerant line to be used. Equipment manufacturers select a valve size or connection fitting on the basis of its application to an average system and other factors, such as the application, length of connecting lines, type of system control, variation in load, and a multitude of other factors. It is entirely possible for the required refrigerant line size to be either smaller or larger than the fittings on various system components. In such cases, reducing fittings must be used.

OIL RETURN

Since oil must pass through the compressor cylinders to provide the proper lubrication, a small amount of oil is always circulating through the system with the refrigerant. The oils used in the refrigeration systems are soluble in liquid refrigerant, and at normal room temperatures they will mix completely. Oil and refrigerant vapor, however, do not mix readily, and the oil can be properly circulated through the system only if the velocity of the refrigerant vapor is great enough to carry the oil along with it. To assure proper oil circulation, adequate refrigerant velocities must be maintained not only in the suction and discharge lines, but in the evaporator circuits as well.

Several factors combine to make oil return most critical at low evaporating temperatures. As the suction pressure decreases and the refrigerant vapor becomes less dense, it becomes more difficult for the refrigerant to carry the oil along. At the same time, as the suction pressure decreases, the compression ratio increases and the compressor capacity is reduced. Therefore, the weight of the refrigerant that is circulated decreases. Refrigeration oil alone becomes as thick as molasses at temperatures below 0°F. As long as it is mixed with a sufficient amount of liquid refrigerant, however, it flows freely. As the percentage of oil in the mixture increases, the viscosity increases.

At low-temperature conditions, several factors start to converge and can create a critical condition. The density of the refrigerant vapor decreases, the velocity decreases, and as a result more oil starts to accumulate in the evaporator. As the oil and refrigerant mixture become thicker, the oil may start logging in the evaporator rather than returning to the compressor. This results in wide variations in the compressor crankcase oil level in poorly designed systems.

Oil logging in the evaporator can be minimized with adequate refrigerant velocities and properly designed evaporators even at extremely low evaporating temperatures. Normally, oil separators are necessary for operation at evaporating temperatures below −50°F, to minimize the amount of oil in circulation.

EQUIVALENT LENGTH OF PIPE

Each valve, fitting, and bend in a refrigerant line contributes to the friction pressure drop because of its interruption or restriction of smooth flow. Because of the detail and complexity of computing the pressure drop of each individual fitting, normal practice is to establish

TABLE 6-1
Equivalent Lengths of Straight Pipe for Valves
and Fittings in Feet (m)

(OD, In.) Line Size (mm.)	Globe Valve	Angle Valve	90° Elbow	45° Elbow	Tee Line	Tee Branch
½ (12.7)	9 (2.74)	5 (1.52)	0.9 (0.27)	0.4 (0.12)	0.6 (0.18)	2.0 (0.61)
⅝ (15.87)	12 (3.66)	6 (1.83)	1.0 (0.3048)	0.5 (0.15)	0.8 (0.24)	2.5 (0.76)
⅞ (22.23)	15 (4.57)	8 (2.44)	1.5 (0.45)	0.7 (0.21)	1.0 (0.3048)	3.5 (1.07)
1⅛ (28.57)	22 (6.70)	12 (3.66)	1.8 (0.55)	0.9 (0.27)	1.5 (0.46)	4.5 (1.37)
1⅜ (34.93)	28 (8.53)	15 (4.57)	2.4 (0.73)	1.2 (0.36)	1.8 (0.55)	6.0 (1.83)
1⅝ (41.27)	35 (10.67)	17 (5.18)	2.8 (0.85)	1.4 (0.43)	2.0 (0.61)	7.0 (2.13)
2⅛ (53.97)	45 (13.72)	22 (6.70)	3.9 (1.19)	1.8 (0.55)	3.0 (0.91)	10.0 (3.05)
2⅝ (66.67)	51 (15.54)	26 (7.92)	4.6 (1.40)	2.2 (0.67)	3.5 (1.07)	12.0 (3.66)
3⅛ (79.37)	65 (19.81)	34 (10.36)	5.5 (1.68)	2.7 (0.82)	4.5 (1.37)	15.0 (4.57)
3⅝ (92.07)	80 (24.38)	40 (12.19)	6.5 (1.98)	3.0 (0.91)	5.0 (1.52)	17.0 (5.18)

an equivalent length of straight pipe. Pressure drop and line sizing tables and charts are normally set up on a basis of a pressure drop per 100 ft of straight pipe, so the use of equivalent lengths allows the data to be used directly (see Table 6-1).

For accurate calculations of pressure drop, the equivalent length for each fitting should be calculated. As a practical matter, an experienced piping designer may be capable of making an accurate overall percentage allowance unless the piping system is extremely complicated. For long runs of piping of 100 ft or greater, an allowance of 20 to 30% of the actual length may be adequate. For short runs of piping, an allowance as high as 50 to 75% or more of the lineal length may be necessary. Judgment and experience are necessary in making a good estimate, and estimates should be checked frequently with actual calculations to ensure reasonable accuracy.

For items such as solenoid valves and pressure-regulating valves, where the pressure drop through the valve is relatively large, data are normally available from the manufacturer's catalog so that items of this nature can be considered independently of lineal length calculations.

PRESSURE DROP TABLES

There are pressure drop tables available which show the combined pressure drop for all refrigerants. Figures 6-1 and 6-2 are for refrigerants R-22 and R-502. Pressure drops in the discharge line, suction

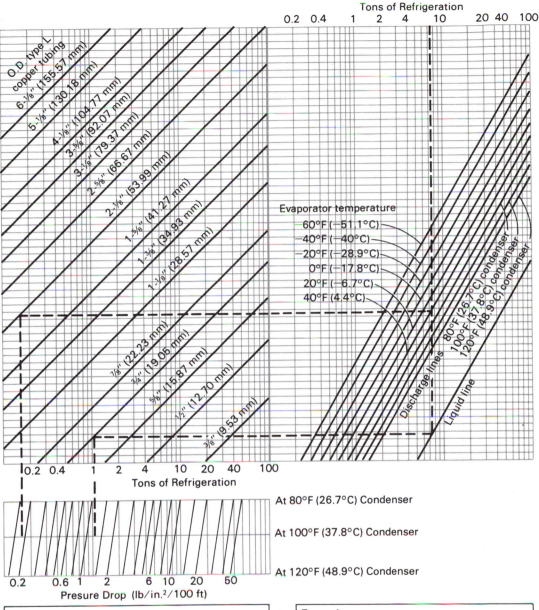

Tons of Refrigeration

0.2 0.4   1   2   4   10   20 40 100

O.D. type L copper tubing

6-¹⁄₈" (155.57 mm)
5-¹⁄₈" (130.18 mm)
4-¹⁄₈" (104.77 mm)
3-⁵⁄₈" (92.07 mm)
3-¹⁄₈" (79.37 mm)
2-⁵⁄₈" (66.67 mm)
2-¹⁄₈" (53.99 mm)
1-⁵⁄₈" (41.27 mm)
1-³⁄₈" (34.93 mm)
1-¹⁄₈" (28.57 mm)

Evaporator temperature
60°F (−51.1°C)
40°F (−40°C)
20°F (−28.9°C)
0°F (−17.8°C)
20°F (−6.7°C)
40°F (4.4°C)

⁷⁄₈" (22.23 mm)
³⁄₄" (19.05 mm)
⁵⁄₈" (15.87 mm)
½" (12.70 mm)
³⁄₈" (9.53 mm)

Discharge lines

80°F (26.7°C) condenser
100°F (37.8°C) condenser
120°F (48.9°C) condenser

Liquid line

0.2 0.4   1   2   4   10 20 40   100
Tons of Refrigeration

At 80°F (26.7°C) Condenser

At 100°F (37.8°C) Condenser

At 120°F (48.9°C) Condenser

0.2   0.6 1   2   6 10   20   50
Presure Drop (lb/in.²/100 ft)

Note:
Pressure drops do not allow for pulsating flow.
If flow is pulsating, use next larger pipe size.
Liquid line determined at 0°F (−17.8°C) evaporator
and 80°F (26.7°C) condenser.
Discharge lines at 0° (−17.8°C) evaporator.
Other conditions do not change results appreciably.
Vapor at evaporator outlet is assumed to be
at 65°F (18.3°C).

Example:
7.5 tons at 40°F (4.44°C) evaporator and
    100°F (37.8°C) condensing temperatures.
2⅛" (53.99 mm) suction line with a pressure
    drop of 0.2 psi/100 ft.
¾" (19.05 mm) liquid line with a pressure
    drop of 1.3 psi/100 ft.

FIGURE 6-1

Refrigerant–22 pressure drop in lines corresponding to a 65°F evaporator outlet.

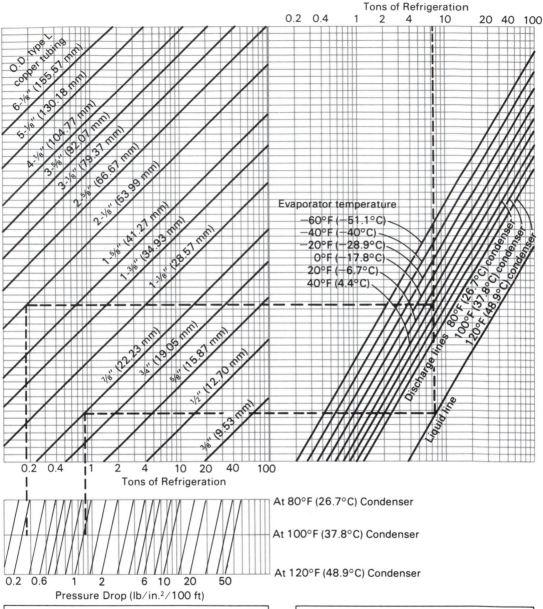

Tons of Refrigeration

Evaporator temperature
- −60°F (−51.1°C)
- −40°F (−40°C)
- −20°F (−28.9°C)
- 0°F (−17.8°C)
- 20°F (−6.7°C)
- 40°F (4.4°C)

Discharge lines
- 80°F (26.7°C) condenser
- 100°F (37.8°C) condenser
- 120°F (48.9°C) condenser

Liquid line

Tons of Refrigeration

At 80°F (26.7°C) Condenser

At 100°F (37.8°C) Condenser

At 120°F (48.9°C) Condenser

Pressure Drop (lb/in.²/100 ft)

Note:
Pressure drops do not allow for pulsating flow.
If flow is pulsating, use next larger pipe size.
Liquid line determined at 0°F (−17.8°C) evaporator and 80°F (26.7°C) condenser.
Discharge lines at 0° (−17.8°C) evaporator.
Other conditions do not change results appreciably.
Vapor at evaporator outlet is assumed to be at 65°F (18.3°C).

Example:
7.5 tons at 40°F (4.44°C) evaporator and 100°F (37.8°C) condensing temperatures.
2⅛″ (53.99 mm) suction line with a pressure drop of 0.2 psi/100 ft.
⅞″ (22.2 mm) liquid line with a pressure drop of 1.2 psi/100 ft.

FIGURE 6-2

Refrigerant–502 pressure drop in lines corresponding to a 65° evaporator outlet.

line, and the liquid line can be determined from these charts for condensing temperatures ranging from 80 to 120°F.

To use the chart, start in the upper right-hand corner with the design capacity. Drop vertically downward on the line representing the operating condition desired. Then move horizontally to the left. A vertical line dropped from the intersection point with each size of copper tubing to the design condensing temperature line allows the pressure drop in pounds per square inch (psi) per 100 ft of tubing to be read directly from the chart. The diagonal pressure drop lines at the bottom of the chart represent the change in pressure drop due to a change in condensing temperature.

For example, in Figure 6-2 for R-502, the dashed line represents a pressure drop determination for a suction line in a system having a design capacity of 7.5 tons or 90,000 Btu/hr operating with an evaporating temperature of 40°F. The 2⅝-in. suction line illustrated has a pressure drop of 0.23 psi per 100 ft at 100°F condensing temperature. The same line with the same capacity would have a pressure drop of 0.19 psi per 100 ft at 80°F condensing temperature, and 0.30 psi per 100 ft at 120°F condensing temperature.

In the same manner, the corresponding pressure drop for any line size and any set of operating conditions within the range of the chart can be determined.

SIZING HOT GAS DISCHARGE LINES

Pressure drop in discharge lines is probably less critical than in any other part of the system. Frequently, the effect on capacity of the discharge line pressure drop is overestimated since it is assumed that the compressor discharge pressure and the condensing pressure are the same. In fact, these are two different pressures. The compressor discharge pressure is greater than the condensing pressure by the amount of the discharge line pressure drop. An increase in pressure drop in the discharge line might increase the compressor discharge pressure materially, but have little effect on the condensing pressure. Although there is a slight increase in the heat of compression for an increase in discharge pressure, the volume of gas pumped is decreased slightly due to a decrease in volumetric efficiency of the compressor. Therefore, the total heat to be dissipated through the condenser may be relatively unchanged, and the condensing temperature and pressure may be quite stable, even though the discharge line pressure drop and, because of the pressure drop, the compressor discharge pressure might vary considerably.

The performance of a typical compressor, operating at air conditioning conditions with R-22 and an air-cooled condenser, indicates that for each 5-psi pressure drop in the discharge line, the compressor capacity is reduced less than ½ of 1%, while the electrical power required is increased about 1%. On a typical low-temperature compressor operating with R-502 and an air-cooled condenser, approximately 1% of the compressor capacity will be lost for each 5-psi pressure drop, but there will be little or no change in electrical power consumption.

As a general guide, for discharge line pressure drops up to 5 psi, the effect on the system performance would be so small that it would be difficult to measure. Pressure drops up to 10 psi would not be greatly detrimental to system performance provided that the condenser is sized to maintain reasonable condensing pressures.

Actually, a reasonable pressure drop in the discharge line is often desirable to dampen compressor discharge pulsation, and thereby reduce noise and vibration. Some discharge line mufflers actually derive much of their efficiency from a pressure drop through the muffler.

Discharge lines on a factory-built condensing unit are not a field problem, but on systems installed in the field with remote condensers, line sizes must be selected to provide proper system performance.

Because of the high temperatures that exist in the discharge line, the oil circulation through both horizontal and vertical lines can be maintained satisfactorily with reasonably low refrigerant velocities. Since oil traveling up a riser usually creeps up the inner surface of the pipe, oil travel in vertical risers is dependent on the velocity of the gas at the tubing wall. The larger the pipe diameter, the greater will be the velocity required at the center of the pipe to maintain a given velocity at the wall surface. Figures 6-3 and 6-4 list the maximum recommended discharge line riser sizes for proper oil return for varying capacities. The variation at different condensing temperature is not great, so the line sizes shown are acceptable on both water-cooled and air-cooled applications.

If horizontal lines are run with a pitch in the direction of refrigerant flow of at least ½ in. in 10 ft, there is normally little problem with oil circulation at lower velocities in horizontal lines. However, because of the relatively low velocities required in vertical discharge lines, it is recommended whenever possible that both horizontal and vertical discharge lines be sized on the same basis.

To illustrate the use of the chart we will assume that a system operating with R-12 at 40°F evaporating temperature has a capacity of 100,000 Btu/hr. The intersection of the capacity and evaporating temperature lines at point X on Figure 6-3 indicates the design condition. Since this is below the 2⅛ in.-OD line, the maximum size that can be used to ensure oil return up a vertical riser is a 1⅝-in.-OD line.

Oil circulation in discharge lines is normally a problem only on systems where large variations in system capacity are encountered. For example, an air conditioning system may have steps of capacity control allowing it to operate during periods of light load at capacities

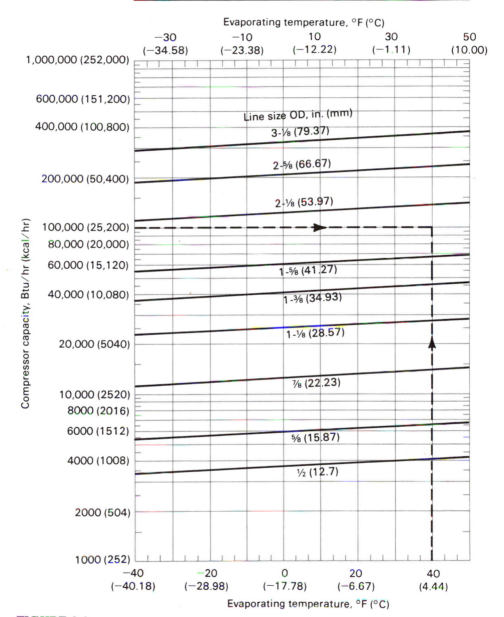

Example:

At 40°F (4.44°C) evaporating temperature and a system capacity of 100,000 Btu/hr (25,200 kcal/hr), maximum size tubing that can be used to insure proper oil return is 1-⅝" (41.27 mm) OD.

Evaporating temperature, °F (°C)

| -30 (-34.58) | -10 (-23.38) | 10 (-12.22) | 30 (-1.11) | 50 (10.00) |

1,000,000 (252,000)

600,000 (151,200)

400,000 (100,800)

Line size OD, in. (mm)

3-⅛ (79.37)

2-⅝ (66.67)

200,000 (50,400)

2-⅛ (53.97)

100,000 (25,200)
80,000 (20,000)

60,000 (15,120)

1-⅝ (41.27)

40,000 (10,080)

1-⅜ (34.93)

1-⅛ (28.57)

20,000 (5040)

⅞ (22.23)

10,000 (2520)
8000 (2016)

6000 (1512)

⅝ (15.87)

4000 (1008)

½ (12.7)

2000 (504)

1000 (252)

Compressor capacity, Btu/hr (kcal/hr)

| -40 (-40.18) | -20 (-28.98) | 0 (-17.78) | 20 (-6.67) | 40 (4.44) |

Evaporating temperature, °F (°C)

FIGURE 6-3

Recommended maximum vertical compressor discharge line sizes to provide proper oil return using R-12 refrigerant.

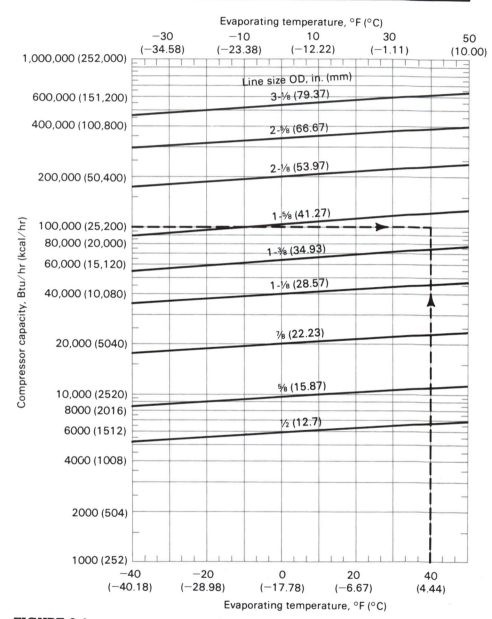

Example:

At 40°F (4.44°C) evaporating temperature and a system capacity of 100,000 Btu/hr (25,200 kcal/hr), maximum size tubing that can be used to insure proper oil return is 1-⅜" (34.93 mm) OD.

Evaporating temperature, °F (°C)

FIGURE 6-4
Recommended maximum vertical compressor discharge line sizes to provide proper oil return using R–22 and R–502 refrigerants.

possibly as low as 25 to 33% of the design capacity. The same situation may exist on commercial refrigeration systems, where compressors connected in parallel are cycled for capacity control. In such cases, vertical discharge lines must be sized to maintain velocities above the minimum velocity necessary to circulate oil properly at the minimum load condition.

For example, consider an air conditioning system using R-12 having a maximum design capacity of 300,000 Btu/hr with steps of capacity reduction up to 66%. Although the 300,000-Btu/hr condition could return oil up to a 2⅝-in.-OD riser at light load conditions, the system would have only 100,000 Btu/hr capacity, so a 1⅝-in.-OD riser must be used. In checking the pressure drop chart (Figure 6-4) at maximum load conditions, a 1⅝-in.-OD. pipe will have a pressure drop of approximately 4 psi per 100 feet at a condensing temperature of 120°F. If the total equivalent length of pipe exceeds 150 ft, in order to keep the total pressure drop within reasonable limits, the horizontal lines should be the next larger size, or 2⅛ in. OD, which would result in a pressure drop of only slightly over 1 psi per 100 ft.

Because of the flexibility in line sizing that the allowable pressure drop makes possible, discharge lines can almost always be sized satisfactorily without the necessity of double risers. If modifications are made to an existing system which result in the existing discharge line being oversized at light load conditions, the addition of an oil separator to minimize oil circulation will normally solve the problem.

One other limiting factor in discharge line sizing is that excessive velocity can cause noise problems. Velocities of 3000 feet per minute (fpm) or more may result in high noise levels, and it is recommended that maximum refrigerant velocities be kept well below this level. Figures 6-5 and 6-6 give equivalent discharge line gas velocities for varying capacities and line sizes over the normal refrigeration and air conditioning range.

SIZING LIQUID LINES

Since liquid refrigerant and oil mix completely, velocity is not essential for oil circulation in the liquid line. The primary concern in liquid line sizing is to ensure a solid liquid column of refrigerant at the expansion valve. If the pressure of the liquid refrigerant falls below its saturation temperature, a portion of the liquid will flash into vapor to cool the remaining liquid refrigerant to the new saturation temperature. This can occur in a liquid line if the pressure drops sufficiently due to line friction or vertical lift.

Flash gas in the liquid line has a detrimental effect on system performance in several ways.

1. It increases the pressure drop due to friction.
2. It reduces the capacity of the flow control devices.
3. It can erode the expansion valve pin and seat.

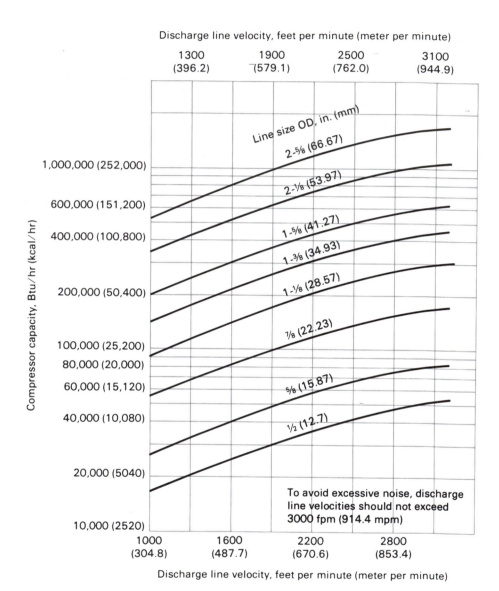

FIGURE 6-5
Recommended discharge line velocities for various Btu/hr (Kcal/hr) capacities using R–22 and R–502 refrigerants.

4. It can cause excessive noise.

5. It can cause erratic feeding of the liquid refrigerant to the evaporator.

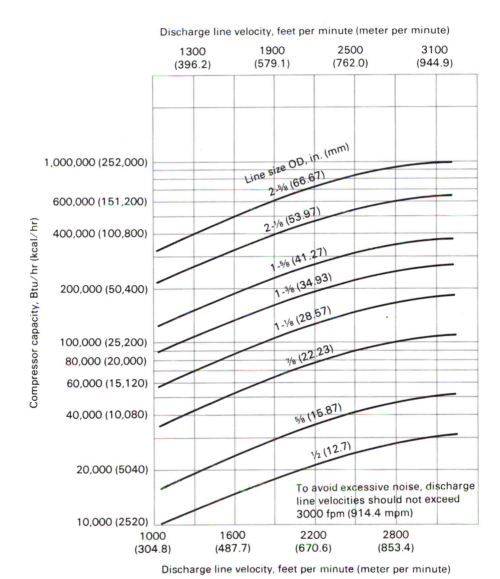

FIGURE 6-6
Recommended discharge line velocities for various Btu/hr (Kcal/hr) capacities using R–12 refrigerant.

For proper system performance, it is essential that liquid refrigerant reaching the flow control device be subcooled slightly below the saturation temperature. On most systems, the liquid refrigerant is sufficiently subcooled as it leaves the condenser to provide for normal system pressure drops. The amount of subcooling necessary, however, depends on the individual system design.

On air-cooled and most water-cooled applications, the temperature of the liquid refrigerant is normally higher than the surrounding ambient temperature, so no heat is transferred into the liquid, and the only concern is the pressure drop in the liquid line. Besides the friction loss caused by the refrigerant flow through the piping, a pressure drop equivalent to the liquid head is involved in forcing liquid refrigerant to flow up to a vertical riser. A vertical head of 2 ft of liquid refrigerant is approximately equivalent to 1 psi. For example, if a condenser or receiver in the basement of a building is to supply liquid refrigerant to an evaporator three floors above, or approximately 30 ft, a pressure drop of approximately 15 psi for the liquid head alone must be provided for in system design.

On evaporative or water-cooled condensers where the condensing temperature is below the ambient air temperature, or on any application where the liquid lines must pass through hot areas such as boiler or furnace rooms, an additional complication may arise because of heat transfer into the liquid. Any subcooling in the condenser may be lost in the receiver or liquid line due to temperature rise alone unless the system is properly designed. On evaporative condensers when a receiver and subcooling coils are used, it is recommended that the refrigerant flow be piped from the condenser to the receiver and then to the subcooling coil. In critical applications, it may be necessary to insulate both the receiver and the liquid line.

On the typical air-cooled condensing unit with a conventional receiver, it is probable that very little subcooling of the liquid is possible unless the receiver is almost completely filled with liquid refrigerant. Vapor in the receiver in contact with the subcooled liquid will condense, and this effect will tend toward a saturated condition.

At normal condensing temperatures, a relation between each 1°F of subcooling and the corresponding change in saturation pressure applies (see Table 6-2).

To illustrate, 5°F subcooling will allow a pressure drop of 8.75 psi with R-12, 13.75 psi with R-22, and 14.25 psi with R-502 without flashing in the liquid line. For the previous example of a condensing unit in a basement requiring a vertical lift of 30 ft, or approximately 15 psi, the necessary subcooling for the liquid head alone would be 8.5°F with R-12, 5.5°F with R-22, and 5.25°F with R-502.

The necessary subcooling may be provided by the condenser used, but for systems with abnormally high vertical risers a suction to the

TABLE 6-2
Relationship of Refrigerant Subcooling and Saturation Pressures

Refrigerant Type	Subcooling	Equivalent Change in Saturation Pressure
R–12	1°F (0.56°C)	1.75 psi (12.06 KN/m²)
R–22	1°F (0.56°C)	2.75 psi (18.96 KN/m²)
R–502	1°F (0.56°C)	2.85 psi (19.65 KN/m²)

liquid heat exchanger may be required. Where long refrigerant lines are involved, and the temperature of the suction gas at the condensing unit is approaching room temperatures, a heat exchanger located near the condenser may not have sufficient temperature differential to cool the liquid adequately. Individual heat exchangers at each evaporator may be necessary.

In extreme cases, where a great deal of subcooling is required, there are several alternatives. A special heat exchanger with a separate subcooling expansion valve can provide maximum cooling with no penalty on system performance. It is also possible to reduce the capacity of the condenser so that a higher operating condensing temperature will make greater subcooling possible. Liquid refrigerant pumps also may be used to overcome large pressure drops.

Liquid line pressure drop causes no direct penalty in electrical power consumption, and the decrease in system capacity due to friction losses in the liquid line is negligible. Because of this, the only real restriction on the amount of liquid line pressure drop is the amount of subcooling available. Most references on pipe sizing recommended a conservative approach with friction pressure drops in the range of 3 to 5 psi. Where adequate subcooling is available, however, many applications have successfully used much higher design pressure drops. The total friction includes line losses through such accessories as solenoid valves, filter-driers, and hand valves.

To minimize the refrigerant charge, liquid lines should be kept as small as practical, and excessively low pressure drops should be avoided. On most systems, a reasonable design criterion is to size liquid lines on the basis of a pressure drop equivalent to 2°F subcooling.

A limitation on liquid line velocity is possible damage to the piping from pressure surges or liquid hammer caused by the rapid closing of liquid line solenoid valves, and velocities above 300 fpm should be avoided when they are used. If liquid line solenoid valves are not used, higher refrigerant velocities can be employed. Figure 6-7 gives liquid line velocities corresponding to various pressure drops and line sizes.

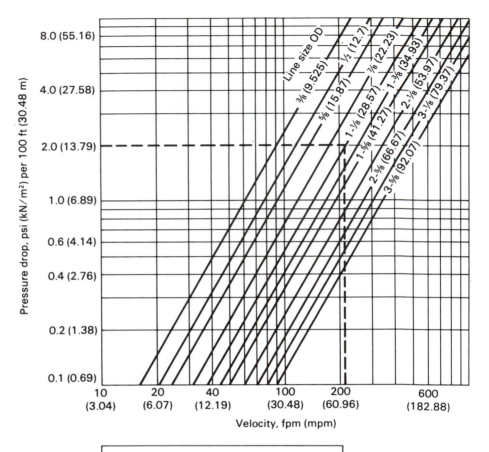

Example:

A pressure drop of 2 psi (13.79 kN/m²) per 100 feet (30.48 m) in a 1-⅛″ (28.57 mm) OD liquid line indicates a velocity of approximately 210 fpm (64 mpm).

FIGURE 6-7
Recommended liquid line velocities for various pressure drops using R–12, R–22, and R–502 refrigerants.

SIZING SUCTION LINES

Suction line sizing is more important than that of the other lines from a design and system standpoint. Any pressure drop occurring due to frictional resistance to flow results in a decrease in the refrigerant pressure at the compressor suction valve, compared with the pressure at the evaporator outlet. As the suction pressure is decreased, each pound of refrigerant that returns to the compressor occupies a greater volume, and the weight of the refrigerant being pumped by the compressor decreases. For example, a typical low-temperature R-502 compressor operating at a −40°F evaporating temperature will lose almost 6% of its rated capacity for each 1 psi of suction line pressure drop.

The normally accepted design practice is to use a design criterion of a suction line pressure drop equal to a 2°F change in saturation temperature. Equivalent pressure drops for various operating conditions are developed and placed in tables (see Table 6-3).

The maintenance of adequate velocities to return the lubricating oil to the compressor properly is also of great importance when sizing suction lines. Studies have shown that oil is most viscous in a system after the suction gas has warmed up a few degrees higher than the evaporating temperature so that the oil is no longer saturated with refrigerant. This condition occurs in the suction line after the refrigerant vapor has left the evaporator. The movement of oil through the suction line is dependent on both the mass and the velocity of the suction vapor. As the mass or density decreases, higher refrigerant velocities are required to force the oil along.

Nominal minimum velocities of 700 fpm in horizontal suction lines and 1500 fpm in vertical suction lines have been recommended and used successfully for many years as suction line sizing design standards. The use of one nominal refrigerant velocity provided a simple and convenient means of checking velocities. However, tests have shown that in vertical risers the oil tends to crawl up the inner surface of the tubing, and the larger the tubing, the greater the velocity required in the center of the tubing to maintain tube surface velocities

TABLE 6-3
Refrigerant Pressure Drop Equivalents for a 2°F
(1.11°C) Change in the Saturation Temperature
at Various Evaporating Temperatures

Evaporating Temperature °F (°C)	Pressure Drop, Psi (KN/m²)		
	R–12	R–22	R–502
45 (7.2)	2.0 (13.79)	3.0 (20.68)	3.3 (22.75)
20 (−6.7)	1.35 (9.31)	2.2 (15.17)	2.4 (16.55)
0 (−17.8)	1.0 (6.89)	1.65 (11.38)	1.85 (12.76)
−20 (−28.9)	0.75 (5.17)	1.15 (7.93)	1.35 (9.31)
−40 (−40)	0.5 (3.45)	0.8 (5.51)	1.0 (6.89)

that will carry the oil. The exact velocity required in vertical suction lines is dependent on both the evaporating temperature and the size of the line. Under varying conditions, the specific velocity required might be either greater or less than 1500 fpm.

For better accuracy in line sizing, revised maximum recommended vertical suction line sizes based on the minimum gas velocities have been calculated and are plotted in chart form for easy use (see Figures 6-8 and 6-9). These revised recommendations supersede previous vertical suction riser recommendations. No change has been made in the 700 fpm minimum velocity recommendation for horizontal lines (see Figures 6-10 and 6-11).

To illustrate, assume again that a system operating with R-12 at a 40°F evaporating temperature has a capacity of 100,000 Btu/hr. On Figure 6-8 the intersection of the evaporating temperature and capacity lines indicate that a 2⅛-in.-OD line will be required for proper oil return in the vertical risers.

Even though the system might have a much greater design capacity, the suction line sizing must be based on the minimum capacity anticipated in operation under light load conditions after allowing for the maximum reduction in capacity from the capacity control, if used.

Since the dual goals of low pressure drop and high velocities are in direct conflict with each other, obviously some compromises must be made in both areas. As a general approach, in suction line design, velocities should be kept as high as possible by sizing lines on the basis of the maximum pressure drop that can be tolerated. In no case, however, should the gas velocity be allowed to fall below the minimum levels necessary to return the oil to the compressor. It is recommended that a tentative selection of suction line sizes be made on the basis of a total pressure drop equivalent to a 2°F change in the saturated evaporating temperature. The final consideration always must be to maintain velocities adequate to return the lubricating oil to the compressor, even if this results in a higher pressure drop than is normally desirable.

DOUBLE RISERS

On systems that are equipped with capacity control compressors, or where tandem or multiple compressors are used with one or more compressors cycled off for capacity control, single-suction line risers may result in either unacceptably high or low gas velocities. A line that is sized properly for light load conditions may have too high a pressure drop at maximum load. If the line is sized on the basis of full-load conditions, the velocities may not be adequate at light load conditions to move the oil through the tubing. On air conditioning appli-

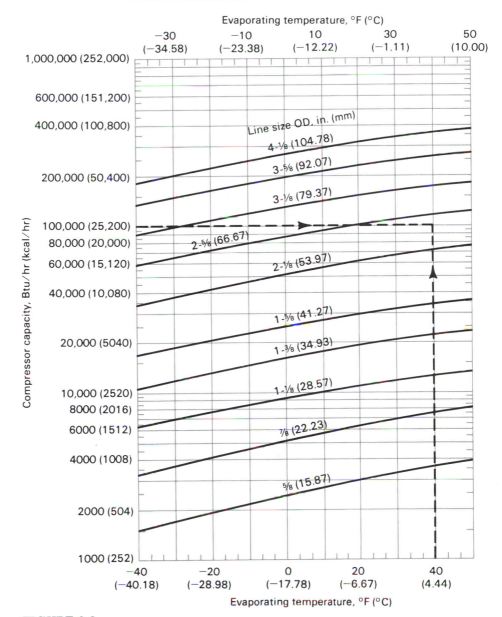

Example:

At 40°F (4.44°C) evaporating temperature and a system capacity of 100,000 Btu/hr (25,200 kcal/hr), the maximum size tubing that can be used to insure proper oil return is 2-⅛" (53.97 mm) OD.

Evaporating temperature, °F (°C)

| −30 (−34.58) | −10 (−23.38) | 10 (−12.22) | 30 (−1.11) | 50 (10.00) |

Line size OD, in. (mm)

4-⅛ (104.78)
3-⅝ (92.07)
3-⅛ (79.37)
2-⅝ (66.67)
2-⅛ (53.97)
1-⅝ (41.27)
1-⅜ (34.93)
1-⅛ (28.57)
⅞ (22.23)
⅝ (15.87)

Compressor capacity, Btu/hr (kcal/hr)

1,000,000 (252,000)
600,000 (151,200)
400,000 (100,800)
200,000 (50,400)
100,000 (25,200)
80,000 (20,000)
60,000 (15,120)
40,000 (10,080)
20,000 (5040)
10,000 (2520)
8000 (2016)
6000 (1512)
4000 (1008)
2000 (504)
1000 (252)

| −40 (−40.18) | −20 (−28.98) | 0 (−17.78) | 20 (−6.67) | 40 (4.44) |

Evaporating temperature, °F (°C)

FIGURE 6-8
Recommended maximum suction line sizes to provide proper oil return using vertical risers and R–12 refrigerant.

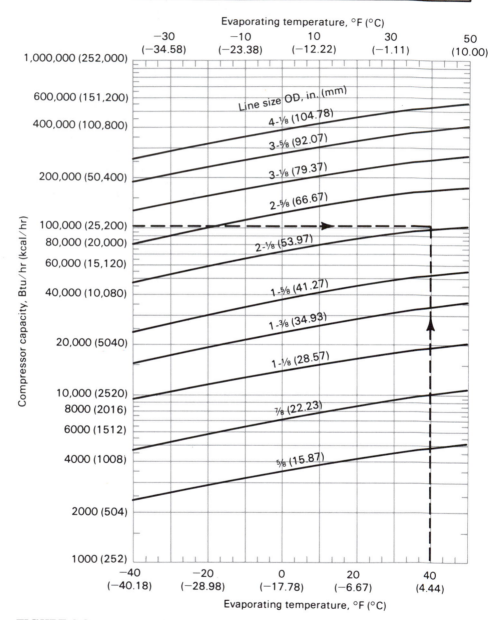

Example:

At 40°F (4.44°C) evaporating temperature and a system capacity of 100,000 Btu/hr (25,200 kcal/hr), the maximum size tubing that can be used to insure proper oil return is 2-⅛" (53.97 mm) OD.

FIGURE 6-9
Recommended maximum suction line sizes to provide proper oil return using vertical risers and R–22 and R–502 refrigerants.

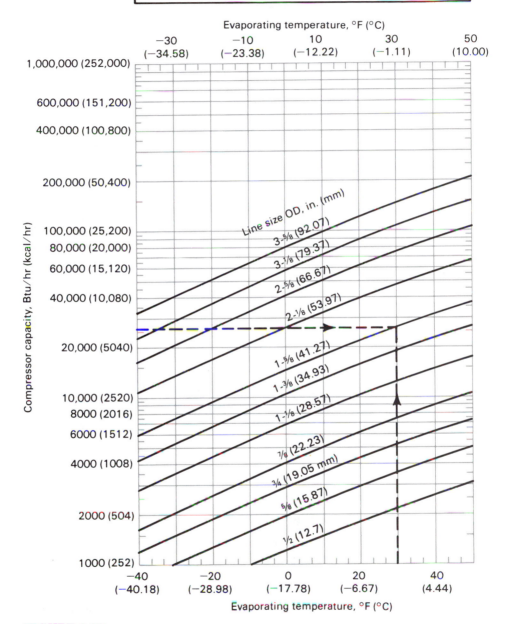

Example: At 30°F (−1.11°C) evaporator temperature, a 1-5⁄8″ (41.27 mm) OD copper suction line should have a unit capacity of not less than 26,000 Btu/hr (6552 kcal/hr) to maintain the desired 700 fpm (213.36 mpm) velocity.

Evaporating temperature, °F (°C)

| −30 (−34.58) | −10 (−23.38) | 10 (−12.22) | 30 (−1.11) | 50 (10.00) |

Compressor capacity, Btu/hr (kcal/hr)

1,000,000 (252,000)
600,000 (151,200)
400,000 (100,800)
200,000 (50,400)
100,000 (25,200)
80,000 (20,000)
60,000 (15,120)
40,000 (10,080)
20,000 (5040)
10,000 (2520)
8000 (2016)
6000 (1512)
4000 (1008)
2000 (504)
1000 (252)

Line size OD, in. (mm)
3-5⁄8 (92.07)
3-1⁄8 (79.37)
2-5⁄8 (66.67)
2-1⁄8 (53.97)
1-5⁄8 (41.27)
1-3⁄8 (34.93)
1-1⁄8 (28.57)
7⁄8 (22.23)
3⁄4 (19.05 mm)
5⁄8 (15.87)
1⁄2 (12.7)

Evaporating temperature, °F (°C)

| −40 (−40.18) | −20 (−28.98) | 0 (−17.78) | 20 (−6.67) | 40 (4.44) |

Evaporating temperature, °F (°C)

FIGURE 6-10
Recommended maximum horizontal suction line sizes to provide proper oil return using R–12 refrigerant.

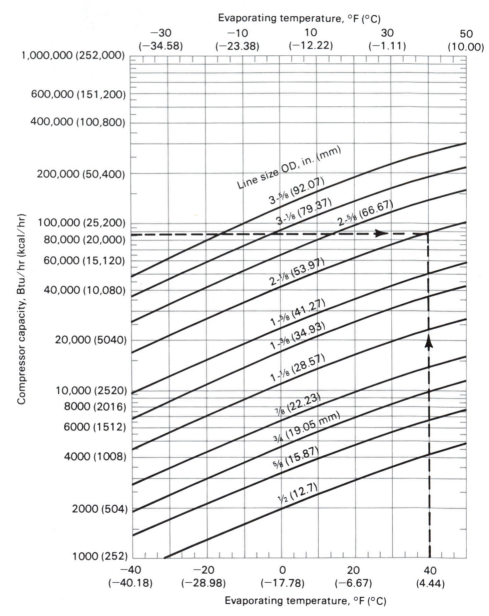

Example: At 40°F (4.44°C) evaporator temperature a 2-⅛″ (53.97 mm) OD copper suction line should have a unit capacity of not less than 86,000 Btu/hr (21,672 kcal/hr) to maintain the desired 700 fpm (213.36 mpm) velocity.

FIGURE 6-11
Recommended maximum horizontal suction line sizes to provide proper oil return using R–22 and R–502 refrigerants.

cations where somewhat higher pressure drops at maximum load conditions can be tolerated without any major penalty in overall system performance, it is usually preferable to accept the additional pressure drop imposed by a single vertical riser. But on medium- or low-temperature applications where pressure drop is more critical, and where separate risers from individual evaporators are not desirable or possible, a double riser may be necessary to avoid an excessive loss of capacity.

The two lines of a double riser should be sized so that the total cross-sectional area is equivalent to the cross-sectional area of a single riser that would have both satisfactory gas velocity and an acceptable pressure drop at maximum load conditions (see Figure 6-12). The two lines are normally of different sizes, with the larger line being trapped as shown. The smaller line must be sized to provide adequate velocities and acceptable pressure drop when the entire minimum load is carried in the smaller riser.

In operation, at maximum load conditions, gas and entrained oil will be flowing through both risers. At minimum load conditions, the gas velocity will not be high enough to carry the oil up both risers. The entrained oil will drop out of the refrigerant gas flow, and accumulate in the P trap, forming a liquid seal. This will force all the flow up the smaller riser, thereby raising the velocity and assuring oil circulation through the system.

As an example, assume a low-temperature system as follows:

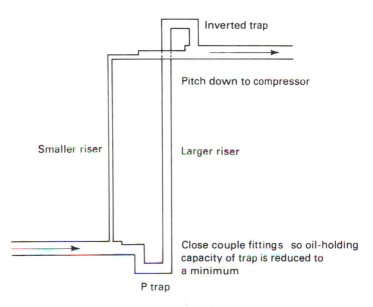

FIGURE 6-12
Refrigerant double riser suction line.

Maximum capacity	150,000
Minimum capacity	50,000
Refrigerant	R-22
Evaporating temperature	$-40°F$
Equivalent length of horizontal piping	125 ft
Vertical riser	25 ft
Desired design pressure drop (equivalent to 2°F)	1 psi

A preliminary check of the R-502 pressure drop chart indicates for a 150-ft run with 150,000 Btu/hr capacity and a total pressure drop of approximately 1 psi, a 3⅛-in.-OD line is indicated (see Figure 6-5). At the minimum capacity of 50,000 Btu/hr a 3⅝-in.-OD horizontal suction line is acceptable (see Figure 6-13). However, Figure 6-12 indicates that a maximum vertical riser size is 2⅛ in. OD. Referring again to the pressure drop chart of Figure 6-5, the pressure drop for 150,000 Btu/hr through 2⅛-in.-OD tubing is 4 psi per 100 ft, or 1 psi for the 25-ft suction riser. Obviously, either a compromise must be made in accepting a greater pressure drop at maximum load conditions, or a double riser must be used.

If the pressure drop must be kept to a minimum, the size of the double riser must be determined. At maximum load conditions, a 3⅛-in.-OD riser would maintain adequate velocities, so a combination of

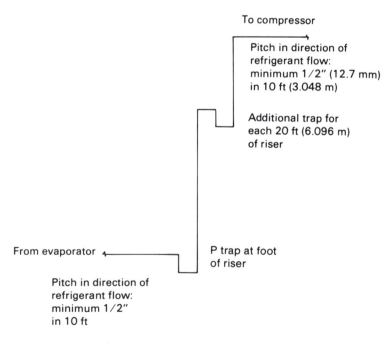

FIGURE 6-13
Refrigerant suction line riser.

line sizes approximating the 3⅛-in.-OD line can be selected for the double riser. The cross-sectional area of the line sizes to be considered are:

3⅛ in.	6.64 sq in.
2⅝ in.	4.77 sq in.
2⅛ in.	3.10 sq in.
1⅝ in.	1.78 sq in.

At the minimum load conditions of 50,000 Btu/hr, the 1⅝-in.-OD line will have a pressure drop of approximately 0.5 psi and will have acceptable velocities, so a combination of 2⅝-in.-OD and 1⅝-in.-OD tubing should be used for the double riser.

In a similar manner, double risers can be calculated for any set of maximum and minimum capacities where single risers may not be satisfactory.

SUCTION PIPING FOR MULTIPLEX SYSTEMS

It is common practice in supermarket applications to operate several cases, each with a liquid line solenoid valve and expansion valve control, from a single compressor. Temperature control of individual cases is normally achieved by means of a thermostat that opens and closes the liquid line solenoid valve as necessary. This type of system, commonly called *multiplexing*, requires careful attention in design to avoid oil return problems and compressor overheating.

Since the cases fed by each liquid line solenoid valve may be controlled individually, and because the load on each case is relatively constant during operation, individual suction lines and risers are normally run from each case, or group of cases, which are controlled by a liquid line solenoid valve for a minimum pressure drop and maximum efficiency in oil return. This provides excellent control as long as the compressor is operating at its design suction pressure. However, there may be periods of light load when most or all of the liquid line solenoid valves are closed. Unless some means of controlling compressor capacity is provided, this can result in the compressor short-cycling or operating at excessively low suction pressures. Either condition can result not only in overheating the compressor, but in reducing the suction pressure to a level where the gas becomes so rarified that it can no longer return oil properly in lines sized for much greater gas density.

Because of the fluctuations in the refrigeration load caused by the closing of the individual liquid line solenoid valves, some means of compressor capacity control must be provided. In addition, the means of capacity control must be such that it will not allow extreme variations in the compressor suction pressure.

Where multiple compressors are used, cycling of the individual compressors provides satisfactory control. Where multiplexing is done with a single compressor, a hot gas bypass system has proven to be the most satisfactory means of capacity reduction. This system allows the compressor to operate continuously at a reasonably constant suction pressure while compressor cooling can be safely controlled by means of a de-superheating expansion valve.

In all cases, the operation of the system under all possible combinations of heavy load, light load, defrost, and compressor capacity must be studied carefully to make certain that operating conditions will be satisfactory.

Close attention must be given to piping design on multiplex systems to avoid oil return problems. The lines must be properly sized so that the minimum velocities necessary to return the oil are maintained in both horizontal and vertical suction lines under minimum load conditions. Bear in mind that although a hot gas bypass maintains the suction pressure at a proper level, the refrigerant vapor being bypassed is not available in the system to aid in returning the oil.

PIPING DESIGN FOR HORIZONTAL AND VERTICAL LINES

Horizontal suction and discharge lines should be pitched downward in the direction of flow to aid in oil drainage, with a downward pitch of at least ½ in. in 10 ft. Refrigerant lines should always be as short and should run as directly as possible.

Piping should be located so that access to system components is not hindered, and so that any components that could possibly require future maintenance are easily accessible. If piping must be run through boiler rooms or other areas where they will be exposed to abnormally high temperatures, it may be necessary to insulate both the suction and liquid lines to prevent excessive heat transfer into the lines.

Every vertical suction riser greater than 3 to 4 ft in height should have a P trap at the base to facilitate oil return up the riser (see Figure 6-13). To avoid the accumulation of large quantities of oil, the trap should be of minimum depth and the horizontal section should be as short as possible.

Prefabricated wrought copper traps are available, or a trap can be made by using two street ells and one regular ell. Traps at the foot of hot gas risers are normally not required because of the easier movement of oil at higher temperatures. However, it is recommended that

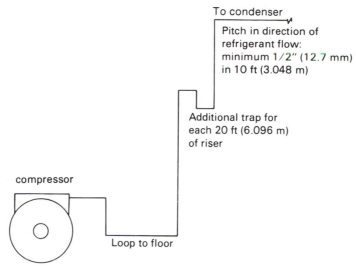

FIGURE 6-14
Compressor discharge line riser.

the discharge line from the compressor be looped to the floor prior to being run vertically upward to prevent the drainage of oil back to the compressor head during shutdown periods (see Figure 6-14).

For long vertical risers in both suction and discharge lines, additional traps are recommended for each full length of pipe (approximately 20 ft) to ensure proper oil movement.

In general, trapped sections of the suction line should be avoided except where necessary for oil return. Oil or liquid refrigerant accumulating in the suction line during the off cycle can return to the compressor at high velocity as a liquid slug on startup, and can break the compressor valves or cause other damage.

SUCTION LINE PIPING DESIGN AT THE EVAPORATOR

If a pumpdown control system is not used, each evaporator must be trapped to prevent liquid refrigerant from draining back to the compressor by gravity during the off cycle. Where multiple evaporators are connected to a common suction line, the connections to the common suction line must be made with inverted traps to prevent drainage from one evaporator from affecting the expansion valve bulb control of another evaporator.

Where a suction riser is taken directly upward from an evaporator,

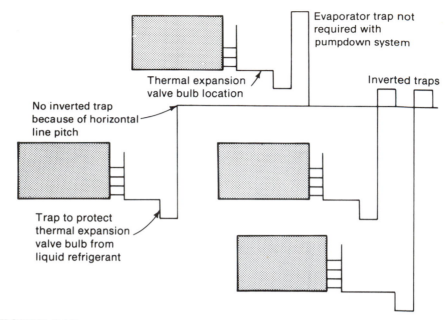

FIGURE 6-15
Typical multiple evaporators suction line piping schematic.

a short horizontal section of tubing and a trap should be provided ahead of the riser so that a suitable mounting for the thermal expansion valve bulb is available. The trap serves as a drain area and helps to prevent the accumulation of liquid under the bulb, which could cause erratic expansion valve operation. If the suction line leaving the evaporator is free-draining, or if a reasonable length of horizontal piping precedes the vertical riser, no trap is required unless necessary for oil return (see Figure 6-15).

SUMMARY

A good refrigerant piping system will have a maximum capacity, be economical, provide proper oil return, provide minimum power consumption, require a minimum amount of refrigerant, have a low noise level, provide proper refrigerant control, and allow perfect flexibility in system performance from 0 to 100% of unit capacity without lubrication problems.

Pressure drop, in general, tends to decrease system capacity and increase the amount of electric power required by the compressor.

It is often required that refrigerant velocity, rather than pressure drop, be the determining factor in system design.

The critical nature of oil return can produce many system difficulties.

An overcharge of refrigerant can result in serious problems of liquid refrigerant control, and the flywheel effect of large quantities of liquid refrigerant in the low-pressure side of the system can result in erratic operation of the refrigerant flow control device.

The size of the refrigerant line connection on a service valve that is supplied with a compressor, or the size of the connection on an evaporator, condenser, or some other system accessory, does not determine the correct size of the refrigerant line to be used.

Oil and refrigerant vapor do not mix readily, and the oil can be properly circulated through the system only if the velocity of the refrigerant vapor is great enough to carry the oil along with it.

To assure proper oil circulation, adequate refrigerant velocities must be maintained not only in the suction and discharge lines, but in the evaporator circuits as well.

Oil logging in the evaporator can be minimized, even at extremely low evaporating temperatures, with adequate refrigerant velocities and properly designed evaporators.

Normally, oil separators are necessary for operation at evaporating temperatures below $-50°F$ in order to minimize the amount of oil in circulation.

Each valve, fitting, and bend in a refrigerant line contributes to the friction pressure drop because of its interruption of restriction of smooth flow.

For accurate calculations of pressure drop, the equivalent length for each fitting should be calculated.

Pressure drop in discharge lines is probably less critical than in any other part of the system.

As a general guide, for discharge line pressure drops up to 5 psi, the effect on the system performance should be so small that it would be difficult to measure.

Actually, a reasonable pressure drop in the discharge line is often desirable to dampen compressor discharge pulsation, and thereby reduce noise and vibration.

Oil circulation in discharge lines is normally a problem only on systems where large variations in system capacity are encountered.

The primary concern in liquid line sizing is to ensure a solid liquid column of refrigerant at the expansion valve.

For proper system performance, it is essential that liquid refrigerant reaching the flow control device be subcooled slightly below its saturation temperature.

Liquid line pressure drop causes no direct penalty in electrical power consumption, and the decrease in system capacity due to friction losses in the liquid line is negligible. Because of this, the only real restriction on the amount of liquid line pressure drop is the amount of subcooling available.

Suction line sizing is more important than that of the other lines from a design and system standpoint. Any pressure drop occurring due to frictional resistance to flow results in a decrease in the refrigerant pressure at the compressor suction valve, compared with the pressure at the evaporator outlet.

The maintenance of adequate velocities to return the lubricating oil to the compressor properly is also of great importance when sizing suction lines.

Nominal minimum velocities of 700 fpm in horizontal suction lines and 1500 fpm in vertical suction lines have been recommended and used successfully for many years as suction line sizing design standards.

As a general approach, in suction line design, velocities should be kept as high as possible by sizing lines on the basis of the maximum pressure drop that can be tolerated. In no case, however, should the vapor velocity be allowed to fall below the minimum levels necessary to return the oil to the compressor.

The two lines of a double riser should be sized so that the total cross-sectional area is equivalent to the cross-sectional area of a single riser that would have both satisfactory vapor velocity and an acceptable pressure drop at maximum load conditions.

Multiplex systems require careful attention in design to avoid oil return problems and compressor overheating.

The lines on multiplex systems must be properly sized so that the minimum velocities necessary to return the oil are maintained in both horizontal and vertical suction lines under minimum load conditions.

Horizontal suction and discharge lines should be pitched downward in the direction of flow to aid in oil drainage, with a downward pitch of at least ½ in. in 10 ft.

Piping should be located so that access to system components is not hindered, and so that any components that could possibly require future maintenance are easily accessible.

Every vertical riser greater than 3 to 4 ft in height should have a P trap at the base to facilitate oil return up the riser.

In general, trapped sections of the suction line should be avoided except where necessary for oil return.

Where multiple evaporators are connected to a common suction line, the connections to the common suction line must be made with inverted traps to prevent drainage from one evaporator from affecting the expansion valve bulb control of another evaporator.

REVIEW QUESTIONS

1. Why should excessive pressure drops in refrigerant lines be avoided?

2. What factor, other than pressure drop, is the determining factor in system design?

3. Is the line connection on a service valve an indication of the correct refrigerant line size?

4. Do oil and refrigerant vapor mix readily?

5. What must be maintained in a refrigeration system to carry oil along with the refrigerant vapor?

6. What happens to the compression ratio as the suction pressure decreases?

7. How can evaporator oil logging be minimized?

8. How do valves, fittings, and bends in a refrigerant line contribute to the friction pressure drop in a refrigeration system?

9. In what line is the pressure drop less critical than in any other part of the system?

10. Which is greater, the compressor discharge or the condensing pressure?

11. In general, what effect on system performance would a pressure drop of up to 5 psi in the discharge line have?

12. Why would a reasonable pressure drop in the discharge line be desirable?

13. In what direction and how much should horizontal refrigerant lines be pitched?

14. Why is oil circulation not usually a problem in the liquid line?

15. What can occur in a liquid line that has too much line friction or vertical lift?

16. How does flash gas affect the flow control device?

17. Does a pressure drop in the liquid line require more electrical power use?

18. How does a pressure drop in the suction line affect the compressor?

19. Why should adequate velocities be maintained in the suction line?

20. What are the nominal minimum velocities recommended in horizontal suction lines?

21. On what type of system are double risers used?

22. On double-riser systems, what should the cross-sectional area of the two lines equal?

23. In piping design on multiplex systems, to what factor should close attention be given?

24. What should be done to refrigerant lines that run through boiler rooms or other areas where they will be exposed to abnormally high temperatures?

25. What should be done to the suction lines when multiple evaporators are connected to a common suction line?

7 Duct Systems and Design

Air distribution—what does it mean to personal comfort? Does it mean a gale of air being blown into a space so strongly that the occupants feel like they are in a whirlwind? Does it mean that there is so little air motion in the space that the occupants feel sleepy and drowsy or stuffy and humid? Actually, it should be neither of these, but some happy medium between them.

INTRODUCTION

Local climatic conditions are a very important factor in proper selection of an air distribution system. There are certain performance characteristics that are needed in certain areas of extreme climatic conditions which are required to maintain indoor comfort on a year-round basis. Areas in cold climates require warm floors in the winter. Hot summer areas require comfort cooling. In areas where both extremes occur, it may be required to have both warm floors and comfort cooling. An air distribution system that performs satisfactorily in one area may not perform properly in another area. A perimeter-type floor

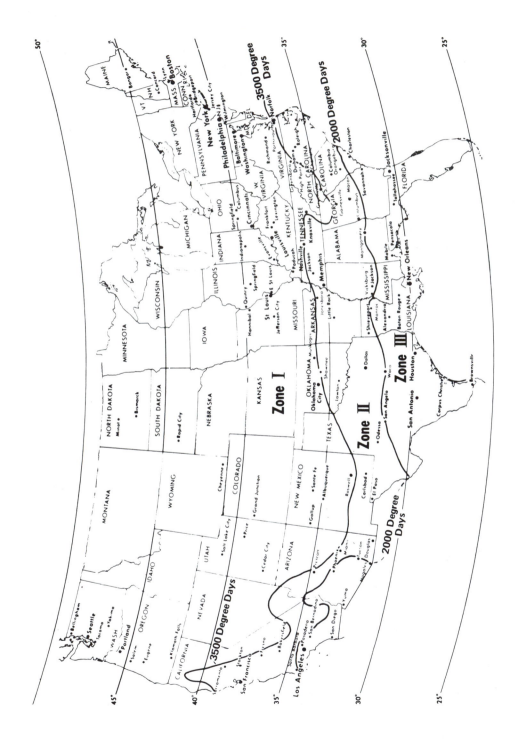

FIGURE 7-1

Degree day zone map. (Courtesy of General Electric Central Air Conditioning Div.)

warming duct system might be recommended in the northern section of the country. At the same time an overhead system without floor warming might be recommended for the same type of building in the South. While the winter and summer inside and outside design conditions determine the capacity requirements of the heating and cooling equipment, it is the overall season conditions that determine the type of air distribution system used.

To aid us in determining what type of air distribution system to select, climatic zones, defined in terms of degree days, have been established (see Figure 7-1). The U.S. continent has been divided into three areas:

Zone I: More than 3500 degree days
Zone II: 2000 to 3500 degree days
Zone III: Fewer than 2000 degree days

Note: The degree day is a unit that represents one degree of difference between the inside temperature and the average outdoor temperature for one day. It is based on the temperature differential below 65°F. For example, on a day when the average outdoor temperature is 45°F, the degree days will be $65 - 45 = 20$. Thus we would have an equivalent of 20 degree days for this one day.

SELECTING AIR DISTRIBUTION SYSTEMS

No single type of air distribution system is best suited for all applications. At times more than one type of system will be used in a single building. Air distribution systems are made up of ducts, supply outlets, and return intakes; therefore, they can be defined in terms of the location of these components.

The location of the supply ducts, supply outlets, and return intakes to maintain proper air circulation and uniform air temperatures should be based on certain factors, such as:

1. The type of residence
   a. Slab floor on the ground
   b. Crawl space beneath the floor
   c. Split level
   d. Multilevel
   e. Apartment
2. Are warm floors a necessity?
3. Will large heat storage in the floors be an advantage or disadvantage?

4. Will zoning be necessary to maintain comfort from top to bottom in split-level or multilevel homes?

5. Is some heat in the crawl space desirable?

6. Is it necessary to heat the full basement?

■ Types of Air Distribution Systems

The *perimeter* system is the most popular and most widely used air distribution system for residential comfort conditioning systems. This system may be defined as a system in which the forced air is conveyed underneath the floor to warm the floor surface, during the heating cycle, and is introduced into the conditioned space near the floor. The air is delivered in an upward direction so as to blanket the outside walls and windows. Indoor comfort is provided with this system by counteracting heat losses or heat gains in the major heat transfer areas.

Perimeter Systems.   There are several types of perimeter air distribution systems. The *loop* system is used to heat slab floors because it provides uniform floor temperatures and satisfactory floor-to-ceiling temperature gradients and a satisfactory room-to-room temperature balance (see Figure 7-2).

The *radial* system is popular in both slab floor and crawl space buildings (see Figure 7-3). This system also provides satisfactory temperature gradients, but with a greater variation in floor temperature.

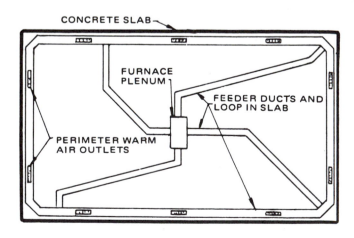

FIGURE 7-2
Perimeter loop system. (Courtesy of General Electric Central Air Conditioning Div.)

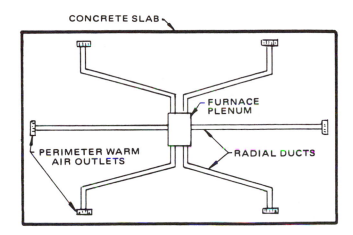

FIGURE 7-3
Perimeter radial system. (Courtesy of General
Electric Central Air Conditioning Div.)

The *trunk and branch* system is used in buildings with basements (see Figure 7-4). It also provides satisfactory temperature gradients.

Ceiling Diffuser System.   In this method of conditioning a home, the forced air is directed horizontal and parallel to the ceiling (see Figure 7-5). During the cooling cycle, air will drift down into the conditioned space. During the heating cycle the air has a tendency to stratify on the ceiling unless it is forced downward into the room. Occupant comfort is not satisfactory when the air does stratify at the ceiling. To aid in air circulation, the return air intakes should be located low on either the inside or outside wall.

High-Inside-Wall Furred Ceiling Systems.   These systems are popular in mild climates (see Figure 7-6). Returns that are direct and close-coupled at the floor level are preferred with these systems.

High-inside-wall systems.   In this method of comfort conditioning the forced air is delivered horizontally across the room toward the outside wall (see Figure 7-7). These systems work well for cooling but are less desirable for heating because of the drafts over the occupants. The return air intakes should be located low on the outside wall to aid in air circulation.

Overhead High-Inside-Wall Systems.   These systems are popular in areas where the summers are hot (see Figure 7-8). Use close-coupled and direct floor-level returns with these systems, especially when the equipment is located within the building.

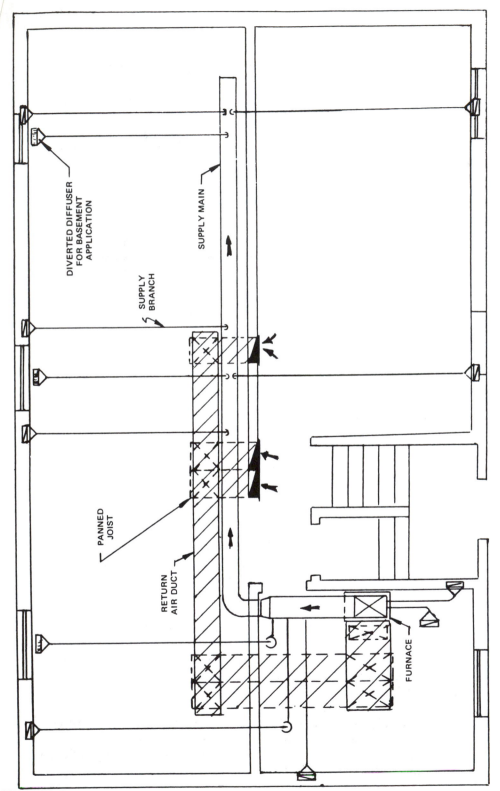

DIVERTED DIFFUSER
FOR BASEMENT
APPLICATION

SUPPLY MAIN

SUPPLY
BRANCH

PANNED
JOIST

RETURN
AIR DUCT

FURNACE

FIGURE 7-4
Perimeter trunk and branch system. (Courtesy of
General Electric Central Air Conditioning Div.)

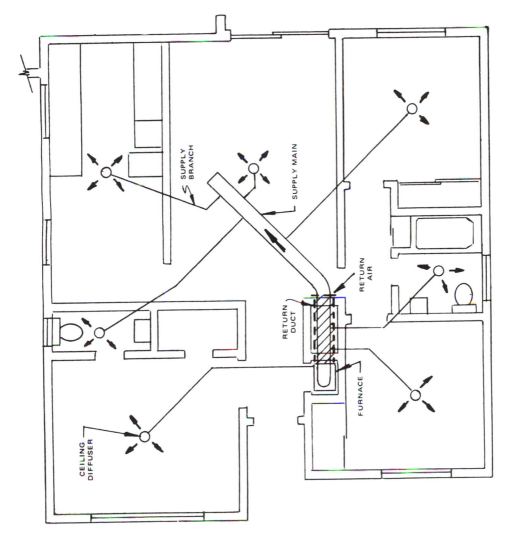

FIGURE 7-5
Ceiling diffuser system. (Courtesy of General
Electric Central Air Conditioning Div.)

171

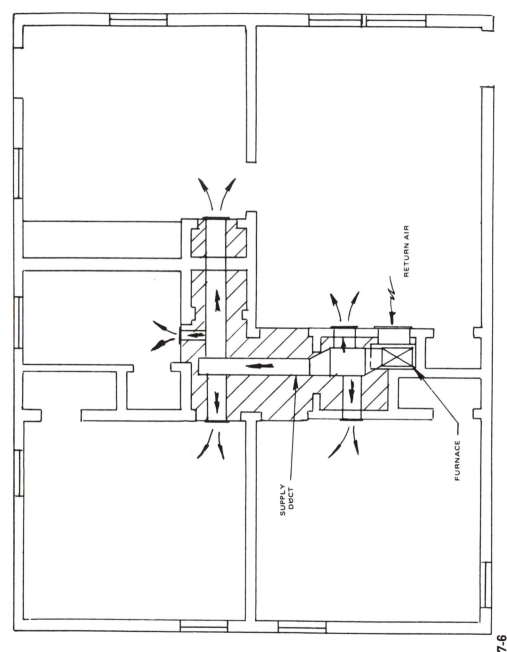

SUPPLY DUCT

RETURN AIR

FURNACE

FIGURE 7-6
High inside wall furred ceiling systems. (Courtesy of General Electric Central Air Conditioning Div.)

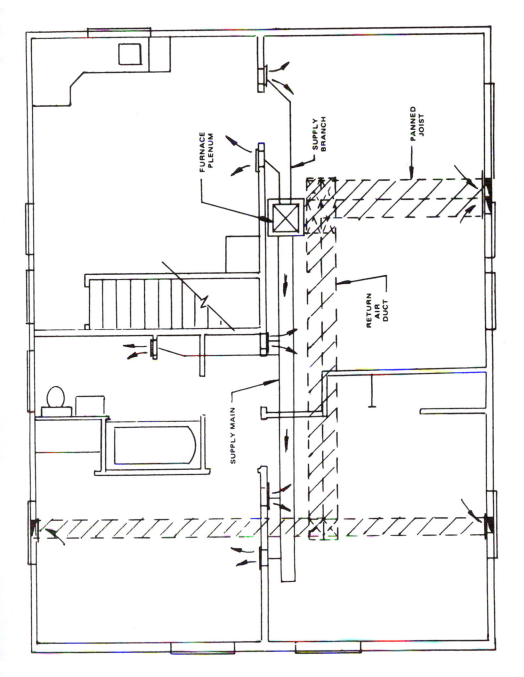

FIGURE 7-7
High inside wall system. (Courtesy of General
Electric Central Air Conditioning Div.)

173

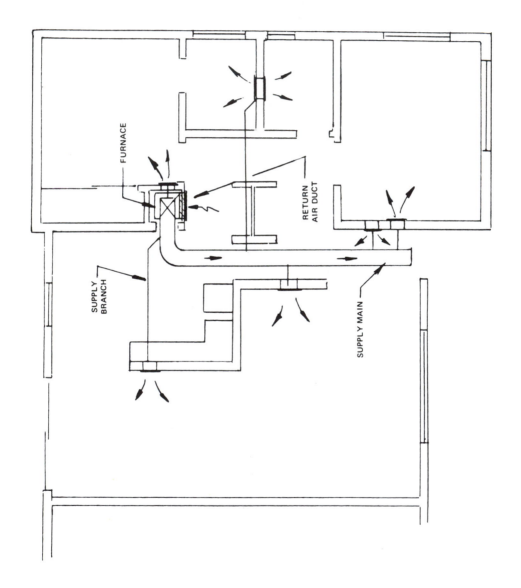

FIGURE 7-8
Overhead high inside wall systems. (Courtesy of
General Electric Central Air Conditioning Div.)

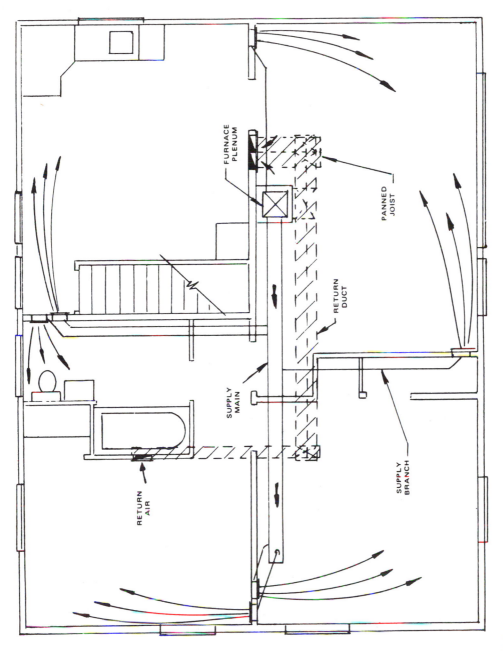

FURNACE PLENUM

PANNED JOIST

RETURN DUCT

SUPPLY MAIN

RETURN AIR

SUPPLY BRANCH

FIGURE 7-9
Parallel flow system. (Courtesy of General Electric Central Air Conditioning Div.)

Parallel-Flow Systems.   These systems may be defined as a method of comfort conditioning where the forced air is directed parallel to an outside wall. The supply outlets are located high on the sidewall next to the outside wall. The air delivery is directed horizontally (see Figure 7-9). The return air intakes should be located on inside walls, but not directly opposite the supply outlet. The return air intakes preferably should be located in the baseboard but can be high in the sidewall. Low supply systems are generally not recommended because it is impossible to avoid drafts on the occupants.

Perimeter System Return Air Intakes.   The return air intakes for perimeter residential air conditioning systems are usually located on the side walls. Even though the return air intakes are located at the baseboard level, the velocity of the air is so low that the air will not cause annoying drafts on the occupants. The return air intakes may be located in each room or provisions must be made to allow the air to pass from all rooms to the central return air intakes.

Listed below are some considerations for return air intakes:

1. Residential return air velocity should be around 500 fpm.

2. They should be located in a stratified zone.

3. They should be located on a wall. Avoid floor or ceiling locations when possible.

4. The grill location does not affect air movement in the room.

■ Special Zoning Considerations

Special considerations should be given to zone control to achieve satisfactory indoor comfort conditions. Temperature variations from room to room should not exceed 3°F. In many types of construction this small variation may be difficult to reach and maintain without a properly controlled zone system.

In split level and trilevel homes, because of the different heat load characteristics in parts of the home, it may be necessary to use zone control for the different levels.

Ranch homes that have large spacious rooms and areas may require zone control in some areas. Also, living areas over unheated spaces present problems in maintaining satisfactory conditions when zone control is not used.

Homes or buildings that are U-, L-, or H-shaped cannot be properly balanced when only one thermostat is used.

Family activity rooms or living areas that are located in basement areas are unique in their heating requirements. The temperatures in such areas are not affected as rapidly by daily weather changes as are the areas above ground and they are greatly affected by the ground temperature. The areas aboveground may require very little heat on a warm sunny winter day while the basement areas may require a good deal of heat because of the losses through the cold basement walls and floor areas.

TYPES OF STRUCTURES

As stated earlier, the air distribution system must be selected on the basis of its performance characteristics, the local climate conditions, and the type of structure. The type of structures and the recommended air distribution systems are listed below.

■ Slab Floor Structures

In Zone I (see Figure 7-1), practically all slab floor buildings must be provided with floor warming if the home is to be comfortable. A perimeter loop or radial system is recommended in this type of climate.

In Zone II, a perimeter system is recommended because of its floor warming characteristics. However, nonperimeter systems that do not provide floor warming are also used.

In Zone III, ceiling diffusers and high-inside-wall supply air outlets are recommended.

■ Crawl Space Structures

In Zone I, perimeter systems are recommended. Either radial or extended plenum systems can be used.

In Zone II, perimeter systems are also recommended. However, nonperimeter systems may be used when assisted with floor warming methods. Systems that use the crawl space as a return air plenum are not acceptable.

In Zone III, all nonperimeter systems are recommended.

■ Basement Structures

In Zone I, perimeter systems are recommended and the basement area must be heated to provide warm floors. Nonperimeter systems are acceptable when equipped with floor warming provisions.

In Zone II, perimeter systems are not recommended. Nonperimeter systems are acceptable without the use of floor warming provisions.

In Zone III, any of the air distributions described previously are recommended. The overhead ceiling or high-wall-supply openings are favored most.

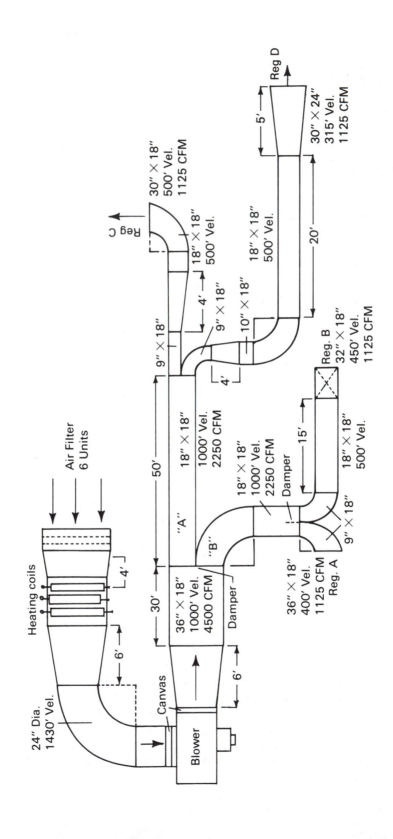

FIGURE 7-10
Supply air system.

■ Split-Level and Multilevel Structures

In Zone I, any slab floors at grade level should use a perimeter system in the floor. Crawl space areas should be equipped with floor warming provisions. Full perimeter systems are generally recommended in this zone.

In Zone II, perimeter systems are generally recommended. However, nonperimeter systems are acceptable.

In Zone III, nonperimeter systems are recommended.

■ Apartment Structures

In Zone I, perimeter systems are generally recommended. Overhead or high wall supply outlets are acceptable for apartments on intermediate floors.

In Zone II, overhead or high-wall-supply outlets are acceptable, especially for intermediate- and top-floor apartments.

In Zone III, overhead or high-wall-supply outlets are recommended.

DUCT DESIGN

When a new air conditioning installation is to be made, it is customary for an engineer to make a layout of the job on paper. This layout, or arrangement, of the job is generally drawn on transparent tracing paper so that blueprints can be made from the drawing or tracing. Drawings may be made in detail where each fitting, duct, or other piece of the mechanism is shown as it actually appears (see Figure 7-10). Notice that all of the fittings and ducts are drawn in considerable detail. The diagram illustrates a typical duct arrangement of an indirect heating and cooling system.

In order to save the time required in drawing or preparing tracings, a set of standard symbols has been adopted. By the use of these symbols the drawings can be correctly interpreted (see Table 7-1). The plan of a heating installation in which the various pipes, fittings, and so on, are indicated by means of symbols instead of being drawn in full is shown in Figure 7-11. To a person who is familiar with the meaning of the symbols, the drawing is entirely clear and a great deal of information is shown on it. A number of special items on the drawings are covered by special notes. For example, the figures around the outside edges of the drawing, followed by a square with the letter "S" drawn through it, indicates in each case the number of square feet of radiation heating surface of the radiator which the pipe serves. The letters "F" and "Q" indicate special fittings—a float vent and a quick vent, respectively—which are not covered by standard symbols.

TABLE 7-1
Ductwork Symbols

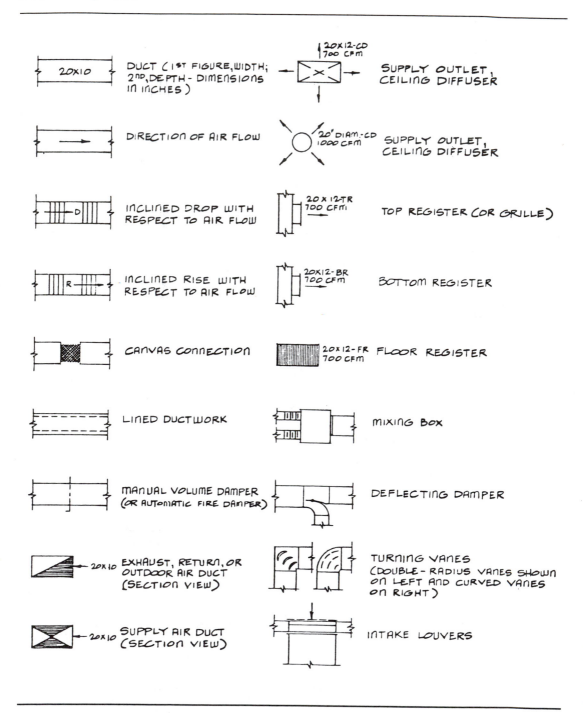

20X10 — DUCT (1ST FIGURE, WIDTH; 2ND, DEPTH - DIMENSIONS IN INCHES)	20X12-CD 700 CFM — SUPPLY OUTLET, CEILING DIFFUSER
DIRECTION OF AIR FLOW	20' DIAM.-CD 1000 CFM — SUPPLY OUTLET, CEILING DIFFUSER
INCLINED DROP WITH RESPECT TO AIR FLOW	20 X 12 TR 700 CFM — TOP REGISTER (OR GRILLE)
INCLINED RISE WITH RESPECT TO AIR FLOW	20X12-BR 700 CFM — BOTTOM REGISTER
CANVAS CONNECTION	20X12-FR 700 CFM — FLOOR REGISTER
LINED DUCTWORK	MIXING BOX
MANUAL VOLUME DAMPER (OR AUTOMATIC FIRE DAMPER)	DEFLECTING DAMPER
20X10 — EXHAUST, RETURN, OR OUTDOOR AIR DUCT (SECTION VIEW)	TURNING VANES (DOUBLE - RADIUS VANES SHOWN ON LEFT AND CURVED VANES ON RIGHT)
20X10 — SUPPLY AIR DUCT (SECTION VIEW)	INTAKE LOUVERS

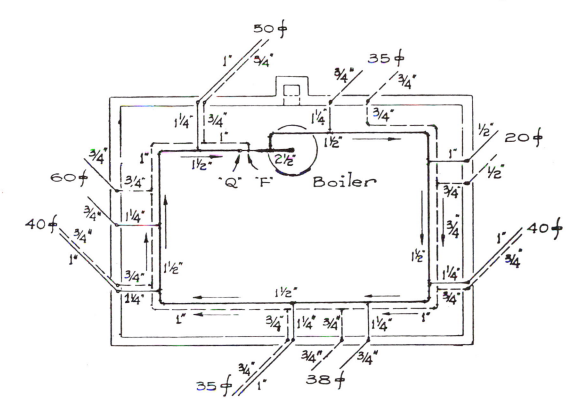

FIGURE 7-11
Piping diagram using symbols.

■ Air Supply

The total quantity of air to be circulated through any building is dependent on the necessity for controlling temperature, humidity, and air distribution when either heating or cooling is required. The factors that determine the total air quantity include the amount of heating or cooling to be done, as well as the type and nature of the building, locality, climate, height of the room, floor area, window area, occupancy, and the method of air distribution.

The air supplied to the conditioned space must always be adequate to satisfy the ventilation requirements of the occupants. It must be sufficient to maintain the desired temperature, humidity, and purity without drafts and with a reasonable degree of uniformity. This total air quantity will ordinarily be composed of two parts: the outdoor or fresh air supply and the recirculated air supply. The requirements as given by the American Society of Heating, Ventilating and Air-Condi-

TABLE 7-2
Minimum Outdoor Air Required for Ventilation
(Subject to Local Code Regulations)

Application	CFM per person
Apartment or residence	10–15
Auditorium	5–7½
Barber shop	10–15
Bank or beauty parlor	7½–10
Broker's board room	25–40
Church	5–7½
Cocktail lounge	20–30
Department store	5–7½
Drugstore	7½–10
Funeral parlor	7½–10
General office space	10–15
Hospital rooms (private)	15–25
Hospital rooms (wards)	10–15
Hotel room	20–30
Night clubs and taverns	15–20
Private office	15–25
Restaurant	12–15
Retail shop	7½–10
Theater (smoking permitted)	10–15
Theater (smoking not permitted)	5–7½

tioning Engineers call for 30 cfm to be circulated per person, at least 10 cfm of which is fresh air.

The minimum amount of fresh air required for ventilation for most common applications has been placed in table form for convenience (see Table 7-2). As should be noted, the figures given in this chart depend to a great extent on the degree of smoking to be expected within the conditioned space. The figures in this table should be used for estimating purposes where they do not conflict with existing local codes. Also, note that this table aims to provide minimum rather than adequate requirements.

■ Duct Construction

Metal ducts are preferable because it is possible to obtain smooth surfaces, thereby avoiding excessive resistance to the airflow. Metal ducts can be also be worked into compact sizes, shapes, and locations.

Metal ducts are usually made either of galvanized sheet iron or aluminum and the sheets are made in various weights or gauges. The

TABLE 7-3
Recommendations for Round Ducts

Diameter (in.)	Gauge of Steel: U.S. Standard	Gauge of Aluminum: B&S
Up to 13	26	24
14–33½	24	24
34–67½	22	20

diameter of the ducts or their width, if they are to be rectangular, determines the gauge of metal (see Table 7-3 for recommended gauges).

All metal duct work should be rigidly constructed and installed to eliminate possible vibration. All slip joints should be in the direction of airflow.

When designing a more extensive system, it is good practice to gradually lower the velocity both in the main duct and the remote branches. This scheme of design has the following advantages:

1. Enables the air to distribute in a uniform way.

2. Decreases the friction in the smaller ducts, where it otherwise would be greatest.

3. When the velocity is lowered, there follows a recovery of velocity pressure, thus compensating for the duct friction.

Where fresh air is supplied, some method should be provided for removing the dust and soot, particularly if the intake is close to the street or alley. Air filters suitable for this purpose are used on the inlet side of the system. Where so used, additional resistance is added and must be allowed for.

Provisions might be made to heat or "temper" the cold incoming air. A blast heating coil may be used. This also adds resistance to the system and must be considered in the calculations.

Dampers and deflectors should be placed at all points necessary to assure proper balance of the system. It is difficult to design a duct system so that the correct amount of air will be delivered at each outlet. Therefore, the use of dampers is essential so that the air supply may be directly proportioned after the system is placed in operation.

Changing Duct Sizes.   In reducing the size or changing the shape of a duct, care must be taken that the angle of the slope is not too abrupt (see Figure 7-12). From this example it is seen that any sharp obstructions in the path of the air through ducts greatly increase the restriction and static pressure. This should be kept in mind when designing a system.

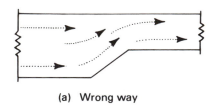
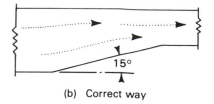

(a) Wrong way   (b) Correct way

FIGURE 7-12
Changing duct sizes.

Elbows. When designing a duct system, great care should be taken in shaping the elbows because sharp turns add greatly to the friction and lower the efficiency of the entire system. Whenever possible, 90° bends should be made with a centerline radius equal to one and one-half times the diameter of the duct at the point of bend, or one diameter at the very least in a tight corner. Figures 7-13 through 7-16 show four different 90° elbows. Figure 7-13 shows a centerline radius of one

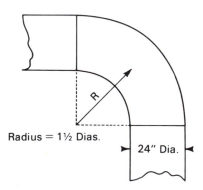

Radius = 1½ Dias.   24" Dia.

FIGURE 7-13
Elbow, radius equal to 1½ diameters.

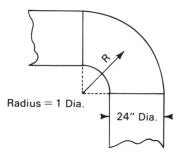

Radius = 1 Dia.   24" Dia.

FIGURE 7-14
Elbow, radius equal to diameter.

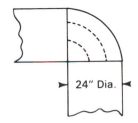

FIGURE 7-15
Elbow, inside throat square, with splitter vanes.

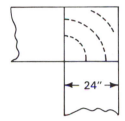

FIGURE 7-16
Elbow, square 90° turn, with turning vanes.

and one-half times the diameter; Figure 7-14, a centerline radius equal to the diameter; Figure 7-15, an inside throat square; and Figure 7-16, a square 90° turn.

A bend similar to Figure 7-13 should be used whenever possible, although a bend similar to Figure 7-14 is permissible where available space will not allow a greater sweep.

Bends similar to Figures 7-15 and 7-16 are not desirable and should be avoided in connection with a duct system, as the power cost is greatly increased. However, these bends can be used when necessary with turning vanes or splitters, which will tend to direct the airstream uniformly around the bends. Without them, congestion and eddy currents would result and increase the resistance to airflow because the entire volume of air would hit the outside of the elbow.

■ Procedure for Duct Design

The general procedure used in duct design is as follows:

1. Do a thorough study of the building plan and locate the supply air outlet positions to provide the proper air distribution within the conditioned space. Choose the outlet sizes from the manufacturer's catalog.

2. Make a sketch of the most convenient duct system. Include both the supply and return ducts from the outlets and intakes to the unit. Make note of the building construction and avoid all obstructions while maintaining a simple design.

3. Calculate the main and branch duct sizes using one of the methods given in the section below headed "Design Methods."

4. Determine the total static pressure requirement of both the supply and return duct systems. Ordinarily, only the pressure loss of the duct run having the greatest static resistance is considered as the pressure loss of the total system, even though the total loss in pressure of each duct run connecting the unit to each supply outlet, or return inlet, should be calculated and made the same for all runs.

5. The more self-balancing the duct system is, the less expensive the overall system is in the long run, from the standpoint of engineering, fabrication, installation, and the balancing of the air flow.

The static pressure regain should be given consideration regardless of the method of design used.

Velocities. By *velocity* is meant the rate of speed of the air traveling through the ducts or openings. The following suggested velocities should be maintained because higher velocities will result in an increase in the noise level and electric power consumption. The velocities

TABLE 7-4
Recommended and Maximum Duct Velocities

Location	Residences	Schools and Public Buildings	Industrial Buildings
Main ducts	700–900 (1000)	1000–1300 (1400)	1200–1800 (2000)
Branch ducts	600 (700)	600–900 (1000)	800–1000 (1200)
Branch risers	500 (650)	600–700 (900)	800 (1000)
Outside air intakes	700 (800)	800 (900)	1000 (1200)
Filters	250 (300)	300 (350)	350 (350)
Heating coils	450 (500)	500 (600)	600 (700)
Air washers	500	500	500
Suction connections	700 (900)	800 (1000)	1000 (1400)
Fan outlets	1000–1600 (1700)	1300–2000 (2200)	1600–2400 (2800)

TABLE 7-5
**Recommended Grill Face Velocities for
Various Applications**

Application	Normal Throw: 10–25 Ft/Min	Long Throw: 30–60 Ft
Radio studios	300–400	Not recommended
Funeral homes	500–600	Not recommended
Residences	500–600	Not recommended
Private offices	650–750	Not recommended
General offices	750–850	1000
Theaters	800–900	1000
Small shops, etc.	800–900	1000
Cafes, bars, etc.	900–1000	1200
Department and grocery stores	900–1200	1500
Industrial plants	1000–1500	1800

given in Tables 7-4, 7-5, and 7-6 are for general purposes. Note that the maximum duct velocities listed in Table 7-4 are given in parentheses. Because the fan horsepower requirement increases as the square of the velocity, approximately, and the noise generated increases with an increase in static pressure, the velocities should be kept low for a quiet and economical operating system. However, it must be remembered that at a given airflow rate, the duct size increases as the velocity decreases. Sometimes it is possible to reduce the between-floor height by using very small ducts in multistory buildings, thus allowing a considerable reduction in the building costs.

■ Design Methods

There are three methods used in designing air conditioning duct systems: (1) equal friction, (2) velocity reduction, and (3) static regain. These three methods represent different design levels of accuracy and

TABLE 7-6
Approximate Velocities for Grills and Registers

Type	Feet per Minute
Baseboard registers	300–500
Wall registers	500–600
Ceiling registers	500–2500
Return grills	500–1000

complexity. Therefore, the method should be selected that will best suit the application. When simple duct systems are used, they may be designed as quickly and as easily as is possible. However, for large installations the system static-pressure requirement must be determined as accurately as possible.

Keep in mind, however, that none of the three design methods listed here will automatically produce the most economical air delivery system for all conditions. If maximum economy is to be reached, a careful evaluation and balancing of all the cost variables that enter into the design of a duct system should be considered with each design method. The main variables that affect the cost of a duct system are ductwork cost, duct insulation, fan horsepower, space requirements, and cost of sound attenuation.

Equal-Friction Method. This method uses the principle of making the pressure loss per foot of length the same throughout the entire system. Very little air balancing is necessary for duct systems in which all the runs have about the same amount of resistance.

This method can be modified, including the design of the longest duct run at two or more different friction values. When a relatively high friction rate must be used on the discharge air side of the system, a lower rate can be used on the return duct system when conditions are less critical to maintain a total system static pressure within the required limits.

Generally, the velocity in the main duct is selected near the fan discharge to provide a satisfactory noise level for a particular installation. Because the flow rate in cfm is generally known, this determines a friction loss per 100 ft of duct (see Figure 7-17 or 7-18). This same amount of friction loss is maintained through the entire system.

The flow rate in the main duct after the first branch takeoff is reduced by the amount taken off by the branch. Thus, in Figures 7-17 and 7-18, proceed vertically downward to the new flow rate value, in cfm, and read the velocity and duct diameter. Notice that the velocity has been reduced. A great advantage of this method is that it automatically reduces the velocities in the ducts in the direction of airflow, thus reducing air noise problems. The equivalent rectangular size for round pipes of any diameter can be found in Table 7-7. By continuing to use this procedure, the design person can size all sections of an air duct system at the same friction loss per linear foot of ductwork.

When the system has been sized, the pressure loss in the duct run which apparently has the greatest resistance should be calculated. The pressure losses caused by all elbows and transitions are to be included and expressed in terms of equivalent length of straight pipe.

The equal-friction method has one main limitation: It does not differentiate between duct runs consisting of several elbows, transitions, and so on, and the runs, which have very little resistance. The actual

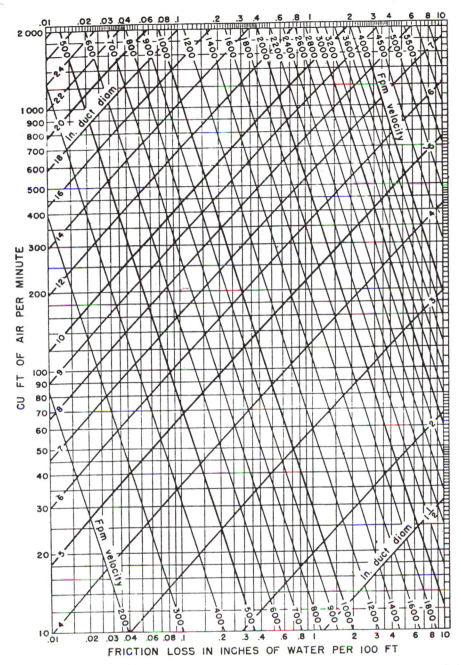

(Based on Standard Air of 0.075 lb per cu ft density flowing through average, clean, round, galvanized metal ducts having approximately 40 joints per 100 ft.) Caution: Do not extrapolate below chart.

FIGURE 7-17
Friction of air in straight ducts for volumes of 10 to 2000 cfm. (Courtesy of American Society of Heating, Refrigerating and Air-Conditioning Engineers, Inc., Atlanta, Ga.)

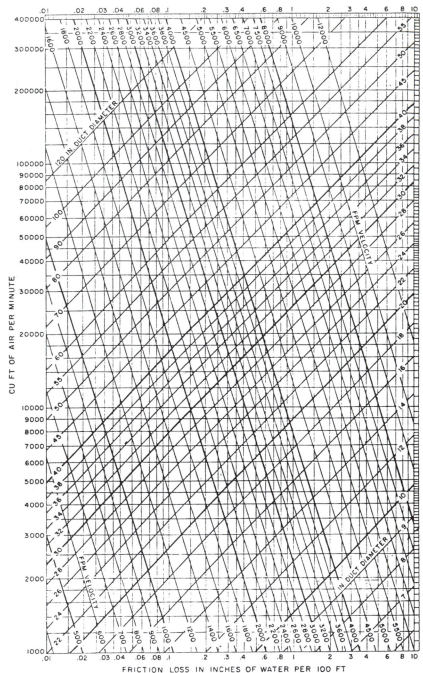

FRICTION LOSS IN INCHES OF WATER PER 100 FT

(Based on Standard Air of 0.075 lb per cu ft density flowing through average, clean, round, galvanized metal ducts having approximately 40 joints per 100 ft.)

FIGURE 7-18
Friction of air in straight ducts for volumes of 1000 to 400,000 cfm. (Courtesy of American Society of Heating, Refrigerating and Air-Conditioning Engineers, Inc., Atlanta, Ga.)

length of duct regulates the cfm flow and determines the duct size. Also, when the system resistance is being computed, the design person must calculate the pressure losses of the various fittings and add these calculations to the straight pipe losses.

When the available pressure for the ductwork is known, this available pressure can be divided by the total equivalent length of the duct run which apparently has the greatest resistance, to determine the friction loss figure per foot. Thus it is not necessary to select an initial airflow velocity for these systems. There is a weakness of this method, however, in that the fitting resistances must be expressed as equivalent length. Transitions, elbows, and so on, have predominately dynamic losses; therefore, the equivalent length of a particular fitting varies considerably with its actual size. The values for elbows are shown in Figure 7-19. This method requires that the duct size be estimated in advance. If there is a considerable difference between the calculated duct size and the initial estimated size, the system should be using the calculated size.

Fewer volume dampers are required if this method is modified so that only the main duct is sized by the equal-friction method. The total duct resistance is used as described for the velocity-reduction method for fan selection. The available air pressure for each branch is divided by its equivalent length in hundreds of feet to calculate a design friction loss figure for use in Figures 7-17 and 7-18, together with the branch airflow in cfm. The branch ducts should be sized as close as possible to use all the available air pressure.

Care should be exercised when using the equal-friction method to prevent branch air velocities from getting too high, thus avoiding noise problems. This problem is easily avoided during the design phase, because the resulting velocity can be read directly on the air friction chart. When the velocity becomes too high, read horizontally to the left-hand column on the friction chart and choose a duct diameter that will provide the desired velocity. The air volume damper for this duct run will be used to reduce the excess pressure. Ductwork attenuates noise to some extent; thus the damper should be located as close to the main duct run as possible. An alternative solution to this type of problem may be to change the duct layout so the resistance on that duct run is increased. Perhaps this can be accomplished by changing the branch takeoff location to increase total duct length.

EXAMPLE 1: A duct layout has three outlets (see Figure 7-20). Outlets 1 and 2 have an airflow of 750 cfm each. Outlet 3 has an airflow of 1000 cfm. The air flows with a velocity of 1600 fpm in section A. Size the duct system and calculate the static-pressure requirement.

Solution: The total cfm to be delivered is $750 + 750 + 1000 = 2500$ cfm. Using Figure 7-18, locate 2500 cfm at 1600-fpm velocity. Read a

TABLE 7-7
Circular Equivalents of Rectangular Ducts for Equal Friction and Capacity

Dimensions in Inches

Side Rectangular Duct	4.0	4.5	5.0	5.5	6.0	6.5	7.0	7.5	8.0	9.0	10.0	11.0	12.0	13.0	14.0	15.0	16.0
3.0	3.8	4.0	4.2	4.4	4.6	4.8	4.9	5.1	5.2	5.5	5.7	6.0	6.2	6.4	6.6	6.8	7.0
3.5	4.1	4.3	4.6	4.8	5.0	5.2	5.3	5.5	5.7	6.0	6.3	6.5	6.8	7.0	7.2	7.4	7.6
4.0	4.4	4.6	4.9	5.1	5.3	5.5	5.7	5.9	6.1	6.4	6.8	7.1	7.3	7.6	7.8	8.1	8.3
4.5	4.6	4.9	5.2	5.4	5.6	5.9	6.1	6.3	6.5	6.9	7.2	7.5	7.8	8.1	8.4	8.6	8.9
5.0	4.9	5.2	5.5	5.7	6.0	6.2	6.4	6.7	6.9	7.3	7.6	8.0	8.3	8.6	8.9	9.1	9.4
5.5	5.1	5.4	5.7	6.0	6.3	6.5	6.8	7.0	7.2	7.6	8.0	8.4	8.7	9.0	9.4	9.6	9.8

Side Rectangular Duct	6	7	8	9	10	11	12	13	14	15	16	17	18	19	20	22	24	26	28	30	Side Rectangular Duct
6	6.6																				6
7	7.1	7.7																			7
8	7.5	8.2	8.8																		8
9	8.0	8.6	9.3	9.9																	9
10	8.4	9.1	9.8	10.4	10.9																10
11	8.8	9.5	10.2	10.8	11.4	12.0															11
12	9.1	9.9	10.7	11.3	11.9	12.5	13.1														12
13	9.5	10.3	11.1	11.8	12.4	13.0	13.6	14.2													13
14	9.8	10.7	11.5	12.2	12.9	13.5	14.2	14.7	15.3												14
15	10.1	11.0	11.8	12.6	13.3	14.0	14.6	15.3	15.8	16.4											15
16	10.4	11.4	12.2	13.0	13.7	14.4	15.1	15.7	16.3	16.9	17.5										16
17	10.7	11.7	12.5	13.4	14.1	14.9	15.5	16.1	16.8	17.4	18.0	18.6									17
18	11.0	11.9	12.9	13.7	14.5	15.3	16.0	16.6	17.3	17.9	18.5	19.1	19.7								18
19	11.2	12.2	13.2	14.1	14.9	15.6	16.4	17.1	17.8	18.4	19.0	19.6	20.2	20.8							19
20	11.5	12.5	13.5	14.4	15.2	15.9	16.8	17.5	18.2	18.8	19.5	20.1	20.7	21.3	21.9						20
22	12.0	13.1	14.1	15.0	15.9	16.7	17.6	18.3	19.1	19.7	20.4	21.0	21.7	22.3	22.9	24.1					22
24	12.4	13.6	14.6	15.6	16.6	17.5	18.3	19.1	19.8	20.6	21.3	21.9	22.6	23.2	23.9	25.1	26.2				24
26	12.8	14.1	15.2	16.2	17.2	18.1	19.0	19.8	20.6	21.4	22.1	22.8	23.5	24.1	24.8	26.1	27.2	28.4			26
28	13.2	14.5	15.6	16.7	17.7	18.7	19.6	20.5	21.3	22.1	22.9	23.6	24.4	25.0	25.7	27.1	28.2	29.5	30.6		28
30	13.6	14.9	16.1	17.2	18.3	19.3	20.2	21.1	22.0	22.9	23.7	24.4	25.2	25.9	26.7	28.0	29.3	30.5	31.6	32.8	30
32	14.0	15.3	16.5	17.7	18.8	19.8	20.8	21.8	22.7	23.6	24.4	25.2	26.0	26.7	27.5	28.9	30.1	31.4	32.6	33.8	32
34	14.4	15.7	17.0	18.2	19.3	20.4	21.4	22.4	23.3	24.2	25.1	25.9	26.7	27.5	28.3	29.7	31.0	32.3	33.6	34.8	34
36	14.7	16.1	17.4	18.6	19.8	20.9	21.9	23.0	23.9	24.8	25.8	26.6	27.4	28.3	29.0	30.5	32.0	33.0	34.6	35.8	36
38	15.0	16.4	17.8	19.0	20.3	21.4	22.5	23.5	24.5	25.4	26.4	27.3	28.1	29.0	29.8	31.4	32.8	34.2	35.5	36.7	38
40	15.3	16.8	18.2	19.4	20.7	21.9	23.0	24.0	25.1	26.0	27.0	27.9	28.8	29.7	30.5	32.1	33.6	35.1	36.4	37.6	40
42	15.6	17.1	18.5	19.8	21.1	22.3	23.4	24.5	25.6	26.6	27.6	28.5	29.4	30.4	31.2	32.8	34.4	35.9	37.3	38.6	42
44	15.9	17.5	18.9	20.2	21.5	22.7	23.9	25.0	26.1	27.2	28.2	29.1	30.0	31.0	31.9	33.5	35.2	36.7	38.1	39.5	44
46	16.2	17.8	19.2	20.6	21.9	23.2	24.3	25.5	26.7	27.7	28.7	29.7	30.6	31.6	32.5	34.2	35.9	37.4	38.9	40.3	46
48	16.5	18.1	19.6	20.9	22.3	23.6	24.8	26.0	27.2	28.2	29.2	30.2	31.2	32.2	33.1	34.9	36.6	38.2	39.7	41.2	48
50	16.8	18.4	19.9	21.3	22.7	24.0	25.2	26.4	27.6	28.7	29.8	30.8	31.8	32.8	33.7	35.5	37.3	38.9	40.4	42.0	50
52	17.0	18.7	20.2	21.6	23.1	24.4	25.6	26.8	28.1	29.2	30.3	31.4	32.4	33.4	34.3	36.2	38.0	39.6	41.2	42.8	52
54	17.3	19.0	20.5	22.0	23.4	24.8	26.1	27.3	28.5	29.7	30.8	31.9	32.9	33.9	34.9	36.8	38.7	40.3	42.0	43.6	54
56	17.6	19.3	20.9	22.4	23.8	25.2	26.5	27.7	28.9	30.1	31.2	32.4	33.4	34.5	35.5	37.4	39.3	41.0	42.7	44.3	56
58	17.8	19.5	21.1	22.7	24.2	25.5	26.9	28.2	29.3	30.5	31.7	32.9	33.9	35.0	36.0	38.0	39.8	41.7	43.4	45.0	58
60	18.1	19.8	21.4	23.0	24.5	25.8	27.3	28.7	29.8	31.0	32.2	33.4	34.5	35.5	36.5	38.6	40.4	42.3	44.0	45.8	60
62	18.3	20.1	21.7	23.3	24.8	26.2	27.6	29.0	30.2	31.4	32.6	33.8	35.0	36.0	37.1	39.2	41.0	42.9	44.7	46.5	62
64	18.6	20.3	22.0	23.6	25.2	26.5	27.9	29.3	30.6	31.8	33.1	34.2	35.5	36.5	37.6	39.7	41.6	43.5	45.4	47.2	64
66	18.8	20.6	22.3	23.9	25.5	26.9	28.3	29.7	31.0	32.2	33.5	34.7	35.9	37.0	38.1	40.2	42.2	44.1	46.0	47.8	66
68	19.0	20.8	22.5	24.2	25.8	27.3	28.7	30.1	31.4	32.6	33.9	35.1	36.3	37.5	38.6	40.7	42.8	44.7	46.6	48.4	68
70	19.2	21.	22.8	24.5	26.1	27.6	29.1	30.4	31.8	33.1	34.3	35.6	36.8	37.9	39.1	41.3	43.3	45.3	47.2	49.0	70
72															39.6	41.8	43.8	45.9	47.8	49.7	72
74															40.0	42.3	44.4	46.4	48.4	50.3	74
76															40.5	42.8	44.9	47.0	49.0	50.8	76
78															40.9	43.3	45.5	47.5	49.5	51.5	78
80															41.3	43.8	46.0	48.0	50.1	52.0	80
82															41.8	44.2	46.4	48.6	50.6	52.6	82
84															42.2	44.6	46.9	49.2	51.1	53.2	84
86															42.6	45.0	47.4	49.6	51.6	53.7	86
88															43.0	45.4	47.9	50.1	52.2	54.3	88
90															43.4	45.9	48.3	50.6	52.8	54.8	90
92															43.8	46.3	48.7	51.1	53.4	55.4	92
96															44.6	47.2	49.5	52.0	54.4	56.3	96

Equation for Circular Equivalent of a Rectangular Duct:[5]

$$d_e = 1.30 \frac{(ab)^{0.625}}{(a+b)^{0.250}} = 1.30 \sqrt[8]{\frac{(ab)^5}{(a+b)^2}}$$

where

a = length of one side of rectangular duct, inches.
b = length of adjacent side of rectangular duct, inches.
d_e = circular equivalent of a rectangular duct for equal friction and capacity, inches.

TABLE 7-7 (Continued)

Dimensions in Inches

Side Rectangular Duct	32	34	36	38	40	42	44	46	48	50	52	56	60	64	68	72	76	80	84	88	Side Rectangular Duct
32	35.0																				32
34	36.0	37.2																			34
36	37.0	38.2	39.4																		36
38	38.0	39.2	40.4	41.6																	38
40	39.0	40.2	41.4	42.6	43.8																40
42	39.9	41.1	42.4	43.6	44.8	45.9															42
44	40.8	42.0	43.4	44.6	45.8	46.9	48.1														44
46	41.7	43.0	44.3	45.6	46.8	47.9	49.1	50.3													46
48	42.6	43.9	45.2	46.5	47.8	48.9	50.2	51.3	52.6												48
50	43.5	44.8	46.1	47.4	48.8	49.8	51.2	52.3	53.6	54.7											50
52	44.3	45.7	47.1	48.3	49.7	50.8	52.2	53.3	54.6	55.8	56.9										52
54	45.0	46.5	48.0	49.2	50.6	51.8	53.2	54.3	55.6	56.8	57.9										54
56	45.8	47.3	48.8	50.1	51.5	52.7	54.1	55.3	56.5	57.8	58.9	61.3									56
58	46.6	48.1	49.6	51.0	52.4	53.7	55.0	56.2	57.5	58.8	60.0	62.3									58
60	47.3	48.9	50.4	51.8	53.3	54.6	55.9	57.1	58.5	59.8	61.0	63.3	65.7								60
62	48.0	49.7	51.2	52.6	54.2	55.5	56.8	58.0	59.4	60.7	62.0	64.3	66.7								62
64	48.7	50.4	52.0	53.4	55.0	56.4	57.7	59.0	60.3	61.6	62.9	65.3	67.7	70.0							64
66	49.5	51.1	52.8	54.2	55.8	57.2	58.6	59.9	61.2	62.5	63.9	66.3	68.7	71.1							66
68	50.2	51.8	53.5	55.0	56.6	58.0	59.5	60.8	62.1	63.4	64.8	67.3	69.7	72.1	74.4						68
70	50.9	52.5	54.2	55.8	57.3	58.8	60.3	61.7	63.0	64.3	65.7	68.3	70.7	73.1	75.4						70
72	51.5	53.2	54.9	56.5	58.0	59.6	61.1	62.6	63.9	65.2	66.6	69.2	71.7	74.1	76.4	78.8					72
74	52.1	53.9	55.6	57.2	58.8	60.4	61.9	63.3	64.8	66.1	67.5	70.1	72.7	75.1	77.4	79.9					74
76	52.7	54.6	56.3	57.9	59.5	61.2	62.7	64.1	65.6	67.0	68.4	71.0	73.6	76.1	78.4	80.9	83.2				76
78	53.3	55.2	57.0	58.6	60.3	62.0	63.4	64.9	66.4	67.9	69.3	71.8	74.5	77.1	79.4	81.8	84.2				78
80	53.9	55.8	57.6	59.3	61.0	62.7	64.1	65.7	67.2	68.7	70.1	72.7	75.4	78.1	80.4	82.8	85.2	87.5			80
82	54.5	56.4	58.2	60.0	61.7	63.4	64.9	66.5	68.0	69.5	71.0	73.6	76.3	79.0	81.4	83.8	86.2	88.6			82
84	55.1	57.0	58.9	60.7	62.4	64.1	65.7	67.3	68.8	70.3	71.8	74.5	77.2	79.9	82.4	84.8	87.2	89.6	91.9		84
86	55.7	57.6	59.5	61.3	63.0	64.8	66.4	68.0	69.5	71.1	72.6	75.4	78.1	80.8	83.3	85.8	88.2	90.6	92.9		86
88	56.3	58.2	60.1	62.0	63.7	65.4	67.0	68.7	70.3	71.8	73.4	76.3	79.0	81.6	84.2	86.8	89.2	91.6	93.9	96.3	88
90	56.9	58.8	60.7	62.6	64.4	66.0	67.8	69.4	71.1	72.6	74.2	77.1	79.9	82.5	85.1	87.8	90.2	92.6	94.9	97.3	90
92	57.4	59.4	61.3	63.2	65.0	66.8	68.5	70.1	71.8	73.3	74.9	77.9	80.8	83.4	86.0	88.7	91.2	93.6	95.9	98.3	92
94	57.9	60.0	61.9	63.8	65.6	67.5	69.2	70.8	72.5	74.1	75.6	78.7	81.7	84.3	86.9	89.6	92.1	94.6	96.9	99.3	94
96	58.4	60.5	62.4	64.4	66.2	68.2	69.8	71.5	73.2	74.3	76.3	79.4	82.6	85.2	87.8	90.5	93.0	95.6	97.9	100.3	96

(Courtesy of American Society of Heating, Refrigerating and Air-Conditioning Engineers, Inc., Atlanta, Ga.)

pipe diameter of 17 in. with a friction loss of 0.2 in. of water column per 100 ft of duct in section A. Now subtract the 750 cfm delivered to outlet 1, to determine that there is 1750 cfm of airflow in section B. Read along the 0.2-in. friction line. All the ducts can be sized now because all the flow rates are known (see Table 7-8).

Velocity-Reduction Method. The velocity-reduction method consists of choosing an air velocity at the fan discharge and designing the ducts for progressively lower air velocities in the main duct at each branch takeoff. For the chosen velocities and the known cfm, the various duct sizes may be obtained directly from Figure 7-17 or 7-18. The equivalent rectangular duct sizes may be obtained from Table 7-7. The static pressure loss of the duct run which apparently has the greatest

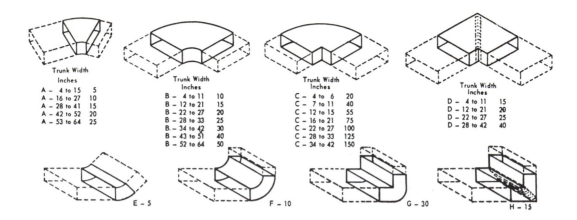

Trunk Width Inches	
A – 4 to 15	5
A – 16 to 27	10
A – 28 to 41	15
A – 42 to 52	20
A – 53 to 64	25

Trunk Width Inches	
B – 4 to 11	10
B – 12 to 21	15
B – 22 to 27	20
B – 28 to 33	25
B – 34 to 42	30
B – 43 to 51	40
B – 52 to 64	50

Trunk Width Inches	
C – 4 to 6	20
C – 7 to 11	40
C – 12 to 15	55
C – 16 to 21	75
C – 22 to 27	100
C – 28 to 33	125
C – 34 to 42	150

Trunk Width Inches	
D – 4 to 11	15
D – 12 to 21	20
D – 22 to 27	25
D – 28 to 42	40

E – 5   F – 10   G – 30   H – 15

(a) Equivalent length of angles and elbows for trunk ducts. (Inside radius = ½ width of duct.)

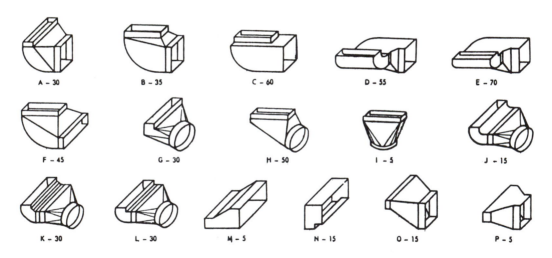

A – 30   B – 35   C – 60   D – 55   E – 70

F – 45   G – 30   H – 50   I – 5   J – 15

K – 30   L – 30   M – 5   N – 15   O – 15   P – 5

(b) Equivalent length of boot fittings. (These values may also be used for floor of diffuser boxes.)

FIGURE 7-19
Equivalent lengths of angles, elbows, and boot fittings. (Courtesy of American Society of Heating, Refrigerating and Air-Conditioning Engineers, Inc., Atlanta, Ga.)

194

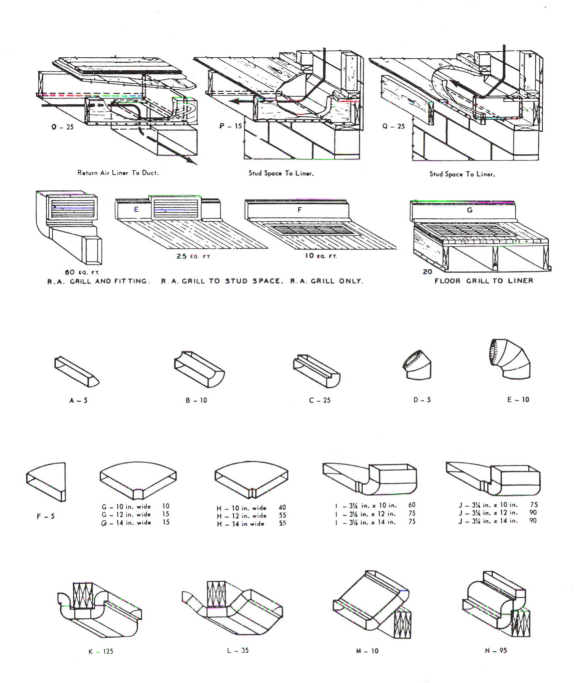

O – 25

Return Air Liner To Duct.

P – 15

Stud Space To Liner.

Q – 25

Stud Space To Liner.

60 EQ. FT.
R.A. GRILL AND FITTING.

25 EQ. FT.
R.A. GRILL TO STUD SPACE.

10 EQ. FT.
R.A. GRILL ONLY.

20
FLOOR GRILL TO LINER

A – 5

B – 10

C – 25

D – 5

E – 10

F – 5

G – 10 in. wide	10
G – 12 in. wide	15
G – 14 in. wide	15

H – 10 in. wide	40
H – 12 in. wide	55
H – 14 in wide	55

I – 3¼ in. x 10 in.	60
I – 3¼ in. x 12 in.	75
I – 3¼ in. x 14 in.	75

J – 3¼ in. x 10 in.	75
J – 3¼ in. x 12 in.	90
J – 3¼ in. x 14 in.	90

K – 125

L – 35

M – 10

N – 95

(c) Equivalent length of angles and elbows for individual and branch ducts. (Inside radius for A and B = 3 in., and for F and G = 5 in.)

FIGURE 7-19 (Continued)

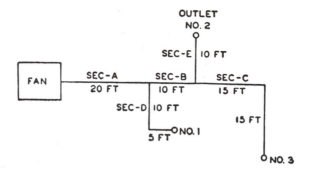

FIGURE 7-20
Duct layout for Example 1. (Courtesy of American
Society of Heating, Refrigerating and Air-Condi-
tioning Engineers, Inc., Atlanta, Ga.)

resistance is found by adding the straight pipe, elbow, and transition losses; this sum is representative of the static pressure required for the supply duct system. The return air duct system is then sized in a similar manner, starting with the lowest air velocities at the return air intakes and increasing them progressively toward the fan inlet. The system is balanced through the use of dampers.

This method may be refined to involve sizing the branch ducts to use the pressure that is available at the entrance to each branch duct. The static pressure loss of the ductwork between the fan and the first branch takeoff is subtracted from the known fan static pressure to determine the available pressure at each takeoff. By using trial and error, a branch velocity is found that will result in the branch pressure loss being equal to, or somewhat less than, that which is available. This procedure is repeated for each branch takeoff.

TABLE 7-8
Tabulation of Results

Section	Flow Rate (Cfm)	Friction per 100 Ft (In H$_2$O)	Duct Diameter (In.)	Velocity (Fpm)	Rectangular Duct (In.)
A	2500	0.2	17.0	1600	20 × 12
B	1750	0.2	14.8	1480	15 × 12
C	1000	0.2	12.0	1290	15 × 8
D	750	0.2	10.7	1190	12 × 8
E	750	0.2	10.7	1190	12 × 8

(Courtesy of American Society of Heating, Refrigerating and Air-Conditioning Engineers, Inc., Atlanta, Ga.)

If the fan is selected to provide a specific static pressure to the ductwork, the method then consists of finding, by trial and error, the main duct velocities that will result in a pressure loss that is equal to the available pressure. The branch ducts are then sized as discussed above.

The advantages of the velocity-reduction design method are: (1) the duct sizes are easily determined, and (2) the velocities can be limited to prevent noise problems. The disadvantages are: (1) the proper choice of air velocities requires experience, and (2) the design person cannot always determine, by inspection, which run will have the greatest amount of resistance.

Static-Regain Method.   In the now popular methods of duct design—equal friction, velocity reduction, or static regain—for average conditions, air velocities in a duct run are progressively reduced, resulting in the conversion of velocity air pressure to static pressure. In these terms, these design methods are static-regain methods. In low-velocity air distribution systems, the conversion of velocity pressure is often disregarded, thus providing a narrow safety margin in the duct system design. However, when considering high-velocity systems, the failure to recognize the amount of static regain will often result in an overdesigned air distribution system with wasted fan motor horsepower and additional air noise problems.

In the ordinary sense, the term *static-regain method* refers to a duct design procedure in which the duct is sized to cause an increase in static pressure or regain at each branch takeoff which will offset any pressure loss because of the succeeding section of the run. When the system or part of the system is designed by the static-regain method only the initial velocity pressure to offset the friction and dynamic losses in the duct system will be required.

This method of designing air distribution systems is especially suited for large, high-velocity systems having several long runs of duct, with each run having branch takeoffs, terminal units, or supply outlets. When this method is used, essentially the same static pressure is available at the entrance to each branch takeoff, outlet, or terminal unit, thus simplifying the selection of outlets or terminals and system balancing.

AIR DISTRIBUTION

The possible comfort conditions in a room are, to a great extent, dependent on the type and location of the supply air outlet grills and to some extent on the location of the return air intake grills.

In general, air supply outlet grills fall into four classifications, which are governed by their discharge air pattern: (1) vertical spreading, (2) vertical nonspreading, (3) horizontal high, and (4) horizontal low. The various grill manufacturers have tables which include the performance of the different types of outlets for both cooling and heating (see Table 7-9). One advantage of the forced air system is that the same grill can be used for both cooling and heating. It should be kept in mind that no single grill is best for both applications.

The best types of outlets for heating use are those providing a vertical spreading air jet located in, or near, the floor next to an outside wall at the point of greatest heat loss, such as under a window. Air delivered at these points will blanket the cold areas and counteract any cold drafts that might occur. This method of air distribution is called perimeter heating.

The best types of outlets for cooling use are located in the ceiling and provide a horizontal discharge air pattern. The air being discharged across the ceiling will blanket this hot area. Also, the air bathes the outside wall to pick up any heat that has entered the building through these areas.

When year-around operation of the system is desired, the type of system chosen depends on the principal application. If heating is to be the major use of the unit, perimeter-type diffusers should be used with the duct system designed to supply maximum supply air velocity for the cooling season. When cooling is to be the major use, ceiling diffusers should be chosen.

The location of return air grills is much more flexible than the supply grill. They may be located in hallways, under windows, in exposed corners, near entrance doors, or on an inside wall, depending on the supply air grill location. Baseboard returns are preferred to floor grills. It is good practice to have them located in or near the outside walls when the supply outlets are located on an inside wall. When a perimeter supply system is used, the return air location is not extremely important because with these systems the return has little effect on the air distribution. Returns that are located centrally on all types of perimeter systems provide satisfactory operation. In larger or multilevel buildings it may be advisable to locate a return air grill in each level or group of rooms. Also, some provision should be made for the return air to move from all the rooms on one level, or group, to the desired return air grill. It is usually desirable to have individual room return air grills to avoid potential problems with airflow under doors or through adjacent rooms and hallways.

When basement-less buildings are to be air conditioned, it is good practice to locate the return air grills either in the ceiling or in the wall. For heating, however, this is simply a matter of preference because returns that are located at the baseboard perform as well as those located in the ceiling.

SIZING A DUCT SYSTEM

When designing a duct system it is often convenient to use a trunk duct, or extended plenum, to supply the feeders (branches) to different rooms.

When sizing a duct system, start at the grill farthest from the equipment and work back to the unit. When there are two or more feeders that are the same distance from the unit, it makes no difference which one is considered farther from the unit.

EXAMPLE 2:   We have determined from a heat load calculation that we need a duct system made up of two 6-in. feeder ducts, one 7-in. duct, and two 8-in. feeders (see Figure 7-21). Size the ducts.

Solution: Either of the two 6-in. feeders may be considered the trunk. When two ducts join, a different size duct results. In this example, when the two 6-in. ducts join, the resulting duct size is 8-in. (see Table 7-10). To read the table, find the trunk size in the left-hand column. We are connecting one 6-in. feeder duct to the 6-in. trunk. Find "1-6" in the center column under "size of feeder joining trunk." In the right-hand column across from the 1-6 entry find 8 in. This is the size of the pipe for section C in the figure. We now have an 8-in. trunk duct picking up a 7-in. feeder (section D). The resulting trunk size is 10-in. (section E). Next, we have a 10-in. trunk picking up an 8-in. feeder (section F). The resulting trunk size is 12-in. (section G). The 12-in. trunk picks up an 8-in. feeder (section H), resulting in a 14-in. trunk duct (section I) back to the unit plenum.

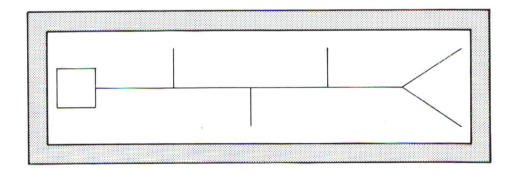

FIGURE 7-21
Round trunk duct system.

TABLE 7-9
Modulaire Series Performance Data—Example of Grill Performance Table

C.F.M.	SIZES	8 × 4 (0)	8 × 4 (22)	8 × 4 (45)	10 × 4 (0)	10 × 4 (22)	10 × 4 (45)	10×5 / 12×4 (0)	(22)	(45)	10×6 / 12×5 / 14×4 / 16×4 (0)	(22)	(45)	12×6 / 14×5 / 18×4 (0)	(22)	(45)	14×6 / 16×5 / 18×5 / 20×4 / 22×4 (0)	(22)	(45)
50	Throw	8	6	5															
	Drop	4.5	2.5	1.0															
	Velocity	335	385	420															
	Static Pressure	.009	.009	.009															
100	Throw	16	13	11	14	12	10	12	11	9	11	10	8	10	9	7	14	12	10
	Drop	6.0	4.0	3.5	6.0	4.5	4.0	7.0	5.0	3.5	7.0	4.5	3.5	6.0	5.0	3.5	8.0	5.5	4.5
	Velocity	670	770	840	530	610	660	440	505	550	375	430	470	290	335	365	390	450	485
	Static Pressure	.028	.037	.044	.017	.023	.027	.012	.016	.019	.009	.011	.013	.009	.009	.009	.009	.013	.015
150	Throw	23	20	17	20	18	15	19	16	13	17	14	12	15	13	11	18	16	14
	Drop	7.0	5.5	4.5	7.0	5.5	4.5	7.5	5.5	4.5	7.5	5.5	4.5	7.5	5.5	4.5	8.5	6.5	5.5
	Velocity	1000	1150	1260	800	915	990	660	760	820	560	645	705	435	500	545	520	600	650
	Static Pressure	.062	.081	.098	.040	.052	.061	.027	.036	.042	.019	.026	.031	.012	.016	.018	.017	.022	.026
200	Throw	31	27		27	24	20	25	22	18	22	19	16	20	18	15	24	20	17
	Drop	8.5	7.0		8.5	7.0	5.5	8.5	6.5	5.5	8.0	6.5	5.0	8.5	7.0	5.5	10.0	7.5	6.0
	Velocity	1340	1540		1060	1220	1320	880	1010	1090	750	860	940	580	670	730	650	750	810
	Static Pressure	.111	.147		.070	.092	.107	.048	.063	.074	.035	.046	.055	.021	.028	.033	.026	.035	.041
250	Throw				35	30		32	28	23	28	24	20	26	22	18	28	24	21
	Drop				10.0	7.5		10.0	7.5	6.0	9.5	7.5	6.0	10.0	7.5	6.0	10.5	8.5	6.5
	Velocity				1330	1525		1100	1265	1360	940	1080	1170	725	833	910	780	900	972
	Static Pressure				.109	.146		.075	.099	.115	.055	.072	.084	.032	.043	.051	.038	.050	.058
300	Throw							38	33		34	29	25	30	26	22	32	28	25
	Drop							11.0	8.5		10.5	8.0	6.5	10.5	8.0	6.5	11.0	8.5	7.5
	Velocity							1320	1420		1120	1290	1405	870	1000	1090	910	1050	1135
	Static Pressure							.107	.144		.077	.104	.121	.047	.062	.074	.051	.068	.078
350	Throw										40	35		36	30	25	38	33	31
	Drop										11.5	9.0		11.5	9.0	7.0	12.5	9.5	8.5
	Velocity										1310	1510		1020	1165	1275	1040	1200	1295
	Static Pressure										.106	.142		.065	.083	.101	.066	.088	.104
400	Throw													41	36	29	42	37	35
	Drop													12.5	10.0	7.5	13.0	10.0	9.0
	Velocity													1160	1330	1455	1170	1350	1460
	Static Pressure													.083	.109	.131	.084	.113	.132
450	Throw													46	40		47	40	
	Drop													13.0	10.5		14.0	10.5	
	Velocity													1310	1500		1300	1500	
	Static Pressure													.106	.140		.105	.140	
500	Throw													50			51		
	Drop													14.0			14.5		
	Velocity													1450			1365		
	Static Pressure													.130			.116		
550	Throw																		
	Drop																		
	Velocity																		
	Static Pressure																		

200

C.F.M.	SIZES	12 × 8 / 16 × 6 / 20 × 5 / 24 × 4			14 × 8 / 18 × 6 / 20 × 6 / 22 × 5 / 28 × 4 / 30 × 4			16 × 8 / 18 × 8 / 22 × 6 / 24 × 6 / 28 × 5 / 30 × 5 / 36 × 4			18 × 10 / 20 × 8 / 22 × 8 / 28 × 6 / 30 × 6 / 48 × 4			18 × 12 / 20 × 10 / 22 × 10 / 24 × 8 / 38 × 6 / 48 × 5			20 × 12 / 22 × 12 / 24 × 10 / 28 × 8 / 30 × 8 / 48 × 6		
		0	22	45	0	22	45	0	22	45	0	22	45	0	22	45	0	22	45
150	Throw	13	11																
	Drop	8.0	6.0																
	Velocity	325	365																
	Static Pressure	.009	.009																
200	Throw	17	15	12	15	14	11	14											
	Drop	9.0	7.0	5.5	9.0	7.5	5.5	9.5											
	Velocity	430	490	535	360	415	450	280											
	Static Pressure	.011	.015	.018	.009	.010	.012	.009											
250	Throw	21	19	15	19	17	14	18	15	12	16								
	Drop	9.5	8.0	6.0	10.0	8.0	5.5	11.0	8.5	6.0	10.5								
	Velocity	540	610	670	450	520	560	350	400	440	300								
	Static Pressure	.018	.023	.028	.012	.017	.019	.009	.010	.012	.009								
300	Throw	26	22	18	23	20	16	21	18	15	19	16	14	18					
	Drop	11.0	8.0	6.0	11.0	8.5	6.0	11.5	9.0	7.0	11.5	9.0	7.0	12.0					
	Velocity	645	735	805	540	625	675	420	480	530	360	415	450	305					
	Static Pressure	.026	.033	.040	.018	.024	.028	.011	.014	.017	.009	.011	.012	.009					
350	Throw	30	26	21	27	24	19	25	22	18	22	19	16	20	18	15			
	Drop	11.5	9.0	6.5	11.5	9.0	7.0	12.5	9.5	7.5	12.5	9.5	7.5	12.5	10.0	7.5			
	Velocity	755	855	940	630	725	785	490	560	615	420	485	525	355	410	450			
	Static Pressure	.035	.045	.055	.025	.033	.038	.015	.019	.023	.011	.014	.017	.009	.010	.012			
400	Throw	35	30	25	31	27	22	28	24	20	26	23	18	24	21	18	22		
	Drop	12.5	9.5	7.5	12.5	9.5	7.0	13.5	10.0	7.5	13.5	10.5	7.5	13.5	11.0	8.0	15.0		
	Velocity	865	975	1075	720	830	895	560	640	705	480	550	600	405	465	525	315		
	Static Pressure	.046	.059	.071	.032	.043	.049	.019	.026	.030	.014	.018	.022	.010	.013	.017	.009		
450	Throw	39	34	28	36	30	25	32	27	22	28	26	21	27	24	20	25	22	17
	Drop	13.5	10.5	8.0	13.5	10.5	8.0	14.5	11.0	8.0	14.0	11.0	8.0	15.0	11.5	8.5	16.0	12.5	9.0
	Velocity	970	1100	1205	810	935	1010	630	720	790	540	620	675	455	525	595	355	405	445
	Static Pressure	.058	.075	.088	.041	.054	.063	.025	.032	.039	.018	.024	.028	.013	.017	.022	.009	.010	.012
500	Throw	42	38	30	39	34	27	35	30	25	32	29	23	30	26	22	28	24	19
	Drop	13.5	11.5	8.0	14.0	10.5	8.0	14.5	11.0	8.5	14.5	11.5	8.5	15.5	12.0	8.5	17.0	12.5	9.5
	Velocity	1080	1220	1340	900	1040	1120	700	800	880	600	690	750	505	585	670	395	455	495
	Static Pressure	.072	.092	.111	.050	.066	.077	.030	.040	.048	.022	.029	.035	.016	.021	.028	.010	.012	.015
550	Throw	46	40	32	41	36	29	37	32	27	34	30	25	32	27	24	30	25	21
	Drop	14.0	11.5	8.5	14.0	11.0	8.0	14.5	11.5	8.5	14.5	11.5	8.5	15.5	12.0	9.0	17.5	12.5	9.5
	Velocity	1190	1345	1475	990	1145	1235	770	880	970	660	760	830	560	640	740	435	500	545
	Static Pressure	.087	.111	.135	.061	.080	.093	.037	.048	.058	.027	.036	.043	.019	.026	.034	.012	.016	.018

(Courtesy of Standard Perforating & Mfg., Inc.)

TABLE 7-10
Diameters of Round Trunks
for Perimeter Loop System

Size of Trunk before Junction (in.)	Size of Feeder Joining Trunk (in.)	Size of Trunk after Junction (in.)
6	1–6	8
	1–7 or 1–8	9
	2–6	9
	2–7	10
	2–8	12
	1–6 & 1–7	10
	1–6 & 1–8	10
	1–7 & 1–8	10
7	1–6 or 1–7	9
	1–8	10
	2–6	10
	2–7 or 2–8	12
	1–6 & 1–7	10
	1–6 & 1–8	12
	1–7 & 1–8	12
8	1–6	9
	1–7 or 1–8	10
	2–6	10
	2–7 or 2–8	12
	1–6 & 1–7	12
	1–6 & 1–8	12
	1–7 & 1–8	12
9	1–6 or 1–7	10
	1–8	12
	2–6 or 2–7	12
	2–8	14
	1–6 & 1–7	12
	1–6 & 1–8	12
	1–7 & 1–8	12

(Courtesy of American Society of Heating, Refrigerating and Air-Conditioning Engineers, Inc., Atlanta, Ga.)

TABLE 7-10 (Continued)

Size of Trunk before Junction (in.)	Size of Feeder Joining Trunk (in.)	Size of Trunk after Junction (in.)
10	1–6, 1–7, or 1–8	12
	2–6 or 2–7 2–8	12 14
	1–6 & 1–7 1–6 & 1–8 1–7 & 1–8	12 12 14
12	1–6, 1–7, or 1–8	14
	2–6 or 2–7 2–8	14 16
	1–6 & 1–7 1–6 & 1–8 1–7 & 1–8	14 14 14
14	1–6, 1–7, or 1–8	16
	2–6, 2–7, or 2–8	16
	1–6 & 1–7 1–6 & 1–8 1–7 & 1–8	16 16 16

TABLE 7-11
Maximum Recommended Velocity

Space	Maximum Velocity
Libraries, Broadcasting Studios, Surgery	500
Residences, Churches, Hotel Bedrooms, Private Offices	750
Banks, Theaters, Cafeterias, School Classrooms, General Office Space, Public Buildings	1000
Factories, Gymnasiums, Warehouses, Arenas, Department Stores	1500

(Courtesy of Standard Perforating & Mfg., Inc.)

DESIGNING A DUCT SYSTEM

Let us design a complete air distribution system for the residence on which we made the heat load calculations in Chapter 3. Since this building is located in Dallas, Texas, the major use of the unit will be for cooling. Therefore, we will need to design the duct system to handle the cooling air volume. We will use rectangular ceiling diffusers in each room (see Figure 7-22).

First, we determine the supply air grill locations. We want the air to be blown toward the point of the greatest heat load. We now need to determine the size of each supply air grill. To do this we must determine the required amount of air for each room. Refer to Figure 7-23. We will use the percentage method for calculating the volume of air in cfm for each room. We will do this by determining what percent of the total heat gain each room represents, then determine the percentage of total air each room represents. The air requirement for the living

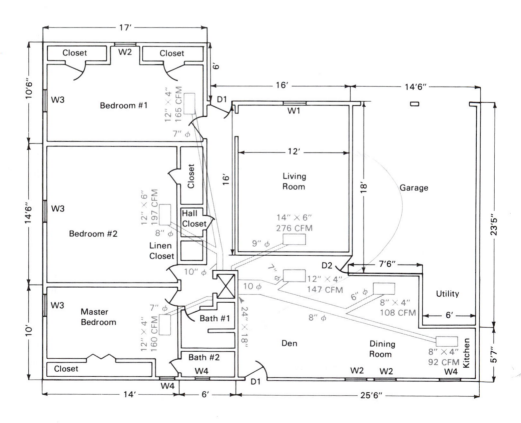

FIGURE 7-22
Model house with duct layout.

room will be: $9672 \div 41,046 \times 100 = 23\%$ of the total heat load. The unit we selected has a rated cfm of 1200. The cfm will be $1200 \times 23\% = 276$ cfm. Enter this figure on line 25 of Figure 7-23 under the "Living Room Cool" heading. Repeat this process for each room and enter the figure in the proper place on the worksheet.

■ Selecting the Proper Supply Grills

The proper selection of the supply air outlet is very important for operation of the air conditioning system. As a part of the total air conditioning system, the air outlet maintains air motion without objectionable drafts in the occupied space.

The air velocities in the conditioned space are a major factor in the comfort of the occupants. Generally, these air velocities should not exceed 50 fpm in the occupied portion of the conditioned space.

Because of the difference in temperature between the supply air and the air in the room, the supply airstream will drop below the level it is being discharged into the room. Caution and good judgment are necessary in the selection and location of the conditioned air supply to prevent the conditioned airstream from entering the occupied zone at a critical velocity which would cause discomfort due to excessive drafts. If the supply air is overthrown and strikes the wall opposite the grill location, objectionable down drafts will occur.

The prescribed general rule is to select a grill with a throw equal to three-fourths the distance to the opposite wall. The terminal velocity at this point should be approximately 6 ft above the floor (see Figure 7-24).

In practice the physical room size and the heating and cooling loads have such wide variation that grills with adjustable deflecting blades are used to help in compensating for these variables.

When applications require spot heating or cooling, grills with a single row of adjustable blades are generally selected. This type of grill will direct the air in one place, either vertical or horizontal, as desired.

Grills that are manufactured with two rows of adjustable blades will direct the air in two planes. This type of grill allows complete flexibility in the adjustment for meeting the throw and drop requirements of the space. These blades are used as a fine adjustment after the best possible selection has been made. If the airstream should prove to be short, the vertical blades may be adjusted to increase the distance of the throw. If the supply airstream enters the occupied zone at too high a velocity, the rear horizontal blades may be adjusted to arch the air above the occupied zone, thus reducing the terminal velocity at this point to help in reducing objectionable drafts.

There are three different settings of the adjustable blades of the supply air grill (0°, 22°, and 45°), which provide different air patterns (see Figure 7-25).

RESIDENTIAL HEATING AND COOLING LOAD ESTIMATE WORKSHEET

	ORIEN-TATION	TD COOL/HEAT	U FACTOR	ENTIRE HOUSE			LIVING ROOM			DINING ROOM			KITCHEN		
					Btu/hr			Btu/hr			Btu/hr			Btu/hr	
				Area	Cool	Heat	Area	Cool	Heat	Area	Cool	Heat	Area	Cool	Heat
WINDOWS															
1	N/S	20/50	0.45				18	162	405	30		675	9		202.5
2															
3	NE/NW														
4															
5	E/W														
6															
7	SE/SW														
8															
WALLS AND PARTITIONS															
9	Wall	20/50	0.18				90.02	324.07	810.18	30		270	44.8	161.28	754.2
10															
11	Part.	20/50	0.21				127.35	534.87	1337.18	104	436.8	1092	48	201.6	504
12															
SUN LOAD															
13	Cost wall	18/50	0.18							30	97.2		39	126.36	
14	East glass	78/50	0.45							30	1053		9	315.9	
15	East door	78/50	0.38												
16															
FLOOR AND CEILING															
17	Ceiling	20/50	0.62				248	3075.2	7688	88	1091.2	2728	33.6	416.64	1041.6
18	Floor	20/50	0.65				248	3224	8060	88	1144	2860	33.6	436.8	1092
DOORS															
19	West	20/50	0.38				19.98	151.85	379.6						
20															
21		Heat loss subtotal				88628.08			18679.98			7625			3594.3
22	Internal cooling load	20% extra						2200						1500	
23	Cooling load	Heat loss		41045.58	88628.08		9671.99	18679.98		3822.2	7625		3158.58	3594.3	
24	% Total load			100			23								
APPROXIMATE AIR QUANTITIES															
25				1200			2761								

FIGURE 7-23
Survey and checklist.

| BATH 1 | | | BEDROOM 1 | | | BEDROOM 2 | | | BEDROOM 3 | | | DEN | | |
Area	Btu/hr Cool	Btu/hr Heat	Area	Btu/hr Cool	Btu/hr Heat	Area	Btu/hr Cool	Btu/hr Heat	Area	Btu/hr Cool	Btu/hr Heat	Area	Btu/hr Cool	Btu/hr Heat
			27	243	607.5	12	108	270	12	108	675			
			241	867.6	2169	104	374.4	936	68	244.8	1890	76.02		684.18
												16.65	69.93	
									142	460.08		76.02	246.3	
									18	631.8				
												19.98	592.2	
60	744	1860	178.5	2213.4	5533.5	246.5	3056.6	7641.5	158	1959.2	4898	162	2008.8	5022
60	780	1950	178.5	2320.5	5801.25	246.5	3204.5	8011.25	158	2054	5135	162	2106	5265
												19.98	379.62	
		3810			14111.25			16858.75			12598			11350.8
	1524	3810		5644.5	14111.25		6743.5	16858.75		5457.88	12598		5022.9	11350.8

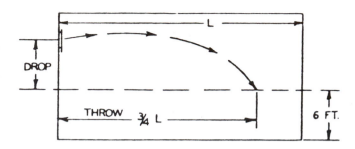

FIGURE 7-24
Recommended length of throw. (Courtesy of
Standard Perforating & Mfg., Inc.)

The amount of noise caused by the supply airstream entering the room through the supply grill is so extremely complicated that manufacturers seldom include decibel ratings in their performance tables. The noise caused by the supply air entering the room is directly proportional to its velocity when leaving the supply grill. The sound level may be kept within satisfactory limits by selecting supply grills with an outlet velocity in recognized approximate ranges (see Table 7-11).

We have determined previously the percentage of the total heat load that each room represents and have determined the volume of air in cfm required to handle this load. These figures should have been entered on the load estimate worksheet.

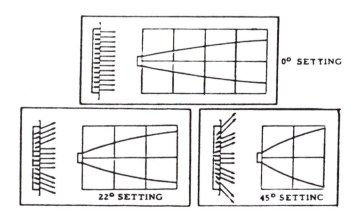

FIGURE 7-25
Different air patterns. (Courtesy of Standard Perforating & Mfg., Inc.)

We can now choose the grills that best fill our needs for the model home. The living room needs 276 cfm to provide the required conditions. We can see on Table 7-9 that no grill is listed that exactly meets our needs. We will choose a cfm of 300. When we follow the numbers indicated for the distance of throw, horizontal line, the size of grill that will deliver the throw nearest our needs is found in the right-hand column. The grill sizes are located at the top of the column. The next step is to choose a grill that is readily available. Thus we will choose a 14 × 6 grill. When the deflector blades are adjusted for 45°, the air throw will be 21 ft with a 6.5-ft drop, a velocity of 972 fpm, and a static pressure of 0.058 in. water column. The throw is a little long and the velocity is a little higher than desired. However, we can make additional adjustments with the volume damper which will cause the grill to perform satisfactorily.

The dining room requires 108 cfm to satisfy its needs. There is not a listing for this amount of cfm. It is much closer to 100 cfm than to 150 cfm; therefore, we will choose an air delivery of 100 cfm on the table. The table lists an 8 × 4 grill which will provide the desired functions. This grill will provide an 11-ft throw, a drop of 3.5 ft, a velocity of 840 fpm, and 0.044 static pressure.

Using this information and examples, the remaining supply air grills can be selected. Write the grill sizes on the house plans beside the grill. As a general rule when the desired air volume is greater than halfway between two listed volumes, choose the next higher listing. If it is less than halfway, choose the lower listing. As can be seen, selecting grills is at best an educated compromise.

■ Selecting the Return Air Grill

The most important points in selecting a return air grill are: (1) it must handle the required cfm, and (2) it must handle the required cfm at the desired velocity. There are tables provided by grill manufacturers to aid in their selection (see Table 7-12).

In our model home we will use a central return under the furnace. The furnace platform is 12 in. above the floor. Thus the maximum height of our return air grill will be 12 in. The unit will move 1200 cfm of air. Checking Table 7-12, the horizontal line along the top of the table represents the air velocity. Since the required air velocity in a residence is 500 fpm, we can follow the 500-fpm column down until we find the allowed volume that is equal to or greater than the 1200 cfm our unit will move. This value is 1265. This grill will handle our unit air satisfactorily because at 500 fpm the volume would be 1265 cfm. Thus the grill size would be 24 × 18. The grill is 24 in. long and 18 in. high. Indicate the grill size at the proper place on the house plan.

TABLE 7-12
Fixed Air Return Grills—HFD, Air Capacities
CFM

SIZE	FREE AREA SQ. FT.	FREE AREA VELOCITY FPM							
		300	400	500	600	700	800	900	1000
10x6	.31	93	124	155	186	217	248	279	310
12x6	.37	111	148	185	222	259	296	333	370
10x8	.41	123	164	205	246	287	328	369	410
12x8	.52	156	208	260	312	364	416	468	520
18x6	.57	171	228	285	342	399	456	513	570
12x12	.80	240	320	400	480	560	640	720	800
18x12	1.23	369	492	615	738	861	984	1107	1230
24x12	1.65	495	660	825	990	1155	1320	1485	1650
18x18	1.87	561	748	935	1122	1309	1496	1683	1870
30x12	2.07	621	828	1035	1242	1449	1656	1863	2070
24x18	2.53	759	1012	1265	1618	1771	2024	2277	2530
30x18	3.19	957	1276	1595	1914	2233	2552	2871	3190
24x24	3.40	1020	1360	1700	2040	2380	2720	3060	3400
36x18	3.84	1152	1536	1920	2304	2688	3072	3456	3840
30x24	4.28	1284	1712	2140	2568	2996	3424	3852	4280
36x24	5.17	1551	2068	2585	3102	3619	4136	4653	5170
36x30	6.56	1968	2624	3280	3936	4592	5248	5904	6560
48x24	6.91	2073	2764	3455	4146	4837	5528	6219	6910
48x30	8.68	2604	3472	4340	5208	6076	6944	7812	8680
48x36	10.50	3150	4200	5250	6300	7350	8400	9450	10500
Press. Drop		.006	.010	.016	.023	.031	.040	.051	.063

(Courtesy of Standard Perforating & Mfg., Inc.)

Now that the grills have been selected and their location determined, we can make a duct layout for the building. This is accomplished by connecting all the supply air grills to the plenum by ductwork (see Figure 7-22). Use the shortest route possible in connecting the ducts. It must be remembered that structural components will dictate the duct route taken. It is usually cheaper and easier to use a trunk duct system where possible, such as feeding the den, dining room, and kitchen with one connection to the discharge plenum, as shown in Figure 7-26.

The next step is to size the ducts to convey the proper amount of air to a given room. Referring to Figures 7-17 and 7-26, we size the ducts to each grill. We will use a 0.1 static pressure on the supply air system. We begin sizing at the grill farthest from the unit in each duct run. Starting with the kitchen, we need 92 cfm. We follow the 0.1 static pressure line, located on the bottom of the table, upward until it intersects with the approximate 92-cfm line, extending from the left side of the table. We find that this is closer to the 6-in. duct than the

5-in. duct; therefore, we will use the 6-in. duct. The duct size for the dining room is found to be almost exactly a 6-in. duct. The den requires a 7-in. duct. These should be indicated on the duct layout (see Figure 7-22). We can now determine the trunk duct size by using Table 7-10. Again, starting with the kitchen, this can be considered the trunk duct. Where the kitchen duct and the dining room duct intersect, we have two 6-in. ducts joining. From the table the resulting trunk duct would be 8-in. in diameter. The result of the 8-in. trunk duct connecting with the 7-in. feeder duct for the den is a 10-in. trunk duct to the discharge air plenum.

The duct to the master bedroom is a 7-in. pipe from the plenum to the supply air grill. There are no feeder ducts in this run.

The remainder of the duct runs can be sized in the same manner. The duct sizes should be indicated on the layout for future reference. Some designers include the cfm of each outlet on the plan. This is for future reference and convenience.

SUMMARY

The local climate conditions are very important in making the proper selection of an air distribution system.

Areas in cold climates require warm floors in the winter. Hot summer areas require comfort cooling.

An air distribution system that performs satisfactorily in one area may not perform properly in another area.

While the winter and summer inside and outside design conditions determine the capacity requirements of the heating and cooling equipment, it is the overall season conditions that determines the type of air distribution system used.

To aid us in finding the type of air distribution system to select, climatic zones, defined in terms of degree days, have been established.

No single type of air distribution system is best suited for all applications.

The location of supply ducts, supply outlets, and return intakes to maintain proper circulation and proper air temperatures should be based on given criteria.

The perimeter system is the most popular and most widely used air distribution system for residential comfort conditioning systems.

The radial system is popular in both slab floor and crawl space buildings.

The trunk and branch system is used in buildings with basements. It also provides satisfactory temperature gradients.

The ceiling diffuser system is popular in buildings with slab floors and where comfort cooling is the major equipment load.

The high-inside-wall furred ceiling system is popular in mild climates.

The overhead high-inside-wall system is popular in areas where the summers are hot.

The return air velocity in a residence should be around 500 FPM.

The return air intake should be located in a stratified zone and on a wall. Avoid floor or ceiling locations when possible.

Temperature variations from room to room should not exceed 3°F.

In split-level and trilevel homes, because of the different heat load characteristics in parts of the home, it may be necessary to use zone control for the different levels.

Homes or buildings that are built in a U, L, or H shape cannot be properly balanced when only one thermostat is used.

The total quantity of air to be circulated through any building is dependent on the necessity for controlling temperature, humidity, and air distribution when either heating or cooling is required.

The total air quantity includes the amount of heating or cooling to be done as well as the type and nature of the building, locality, climate, height of the room, floor area, window area, occupancy, and method of air distribution.

The air supplied to the conditioned space must always be adequate to satisfy the ventilation requirements of the occupants.

Metal ducts are preferable because it is possible to obtain smooth surfaces, thereby avoiding excessive resistance to the airflow.

When designing a more extensive system, it is good practice to gradually lower the velocity both in the main duct and in the remote branches.

Dampers and deflectors should be placed at all points necessary to assure a proper balance of the system.

In reducing the size or changing the shape of a duct, care must be taken that the angle of the slope is not too abrupt.

When designing a duct system, great care should be taken in shaping the elbows because sharp turns add greatly to the friction and lower the efficiency of the entire system.

By velocity is meant the rate of speed of the air traveling through the ducts or openings.

There are three methods used in designing air conditioning duct systems: (1) equal friction, (2) velocity reduction, and (3) static regain.

If maximum economy is to be reached, a careful evaluation and balancing of all the cost variables that enter into the design of a duct system should be considered with each design method.

Care should be exercised when using the equal-friction method to prevent branch air velocities from getting too high, thus avoiding noise problems.

The possible comfort conditions in a room are, to a great extent, dependent on the type and location of the supply air outlet grills and to some extent on the location of the return air intake grills.

The best types of outlets for heating use are those providing a vertical spreading air jet located in, or near, the floor next to an outside wall at the point of greatest heat loss, such as under a window.

The best types of outlets for cooling use are located in the ceiling and provide a horizontal discharge air pattern.

The location of the return air grill is much more flexible than the supply grill.

When designing a duct system it is often convenient to use a trunk duct, or extended plenum, to supply the feeders (branches) to different rooms.

When applications require spot heating or cooling, grills with a single row of adjustable blades are generally selected.

Grills that are manufactured with two rows of adjustable blades will direct the air in two planes.

The noise caused by the supply air entering the room is directly proportional to its velocity when leaving the supply grill.

The most important points in selecting a return air grill are: (1) it must handle the required cfm, and (2) it must handle the required cfm and the desired velocity.

REVIEW QUESTIONS

1. Define a degree day.
2. Upon what is a degree day based?
3. What single type of air distribution system is best suited for all applications?
4. In what direction does the air flow from a ceiling diffuser?
5. List four considerations that should be given to residential return air intakes.
6. Why is zone control sometimes desirable in split-level and trilevel homes?
7. On what does the total quantity of air to be circulated through any building depend?
8. Why is metal preferred in duct construction?
9. Why must abrupt changes in an air duct size be avoided?
10. What is the first step in designing an air distribution system?
11. Define air velocity.

12. What are the three design methods used in designing air distribution systems?

13. Name the four general types of supply air outlet grills.

14. Which is more flexible in its location, the supply grill or the return grill?

15. When sizing a duct system, what point should be considered first?

16. What is the purpose of a supply air outlet grill?

17. What should be the maximum velocity of air in the occupied portion of the conditioned space?

18. To what is the noise caused by the supply air entering the room directly proportional?

19. We are going to connect a 6-in. branch duct to an 8-in. trunk duct. What will be the size of the resulting trunk duct?

20. We have a single room requiring 200 cfm. The duct system will be designed on 0.1 static pressure. What is the required duct size?

Duct Pipe, Fittings, and Insulation

8

To direct the conditioned air to the desired space, some type of carrier is needed. These air carriers are referred to as ducts. They are made of many different types of materials which are fire resistant.

These air carriers work on the principle of air pressure differential. When a difference in pressure is present, air will move from the place of high pressure to the low-pressure area. When this pressure difference is great, the airflow will be faster than when the pressure areas are nearer the same pressure.

The types of duct systems that we are concerned with in most air conditioning installations are low pressure. A low-pressure duct system is one that conveys air at velocities less than 2000 fpm and a static pressure in the duct of 2 in. of water column or less.

INTRODUCTION

Duct systems can be put into two general classifications when used for ventilation: (1) systems where the movement of air is of prime importance, without regard to quiet operation or power economy. In these systems the air velocity will be high. (2) Systems where air must be moved quietly and with power economy. In these systems the air velocity will be low.

Pressure losses in duct systems are caused by the velocity of the airflow, number of fittings, and the friction of the air against the sides of the duct. The use of heating coils, air filters, air washers, dampers, and deflectors all increase the resistance to airflow.

DUCT PIPE SHAPES

Duct pipe is available in round, rectangular, or square shapes. Round duct is generally preferred because it can carry more air in less space. Also, less material is used, and therefore there is less surface to reduce the amount of friction and heat transfer through the duct, and less insulation is required when compared to the other popular shapes. The rectangular shape is generally used where appearance is a factor because the flat surface is easier to work with in regard to the surface of the room or space. Rectangular duct is also sometimes preferred when space is limited or of unusual shape, such as between floor joist or wall studs (see Figure 8-1).

In practice, rectangular duct is used for the plenums and round duct is used for the branch runs and usually the trunk line. Usually, a combination of rectangular duct and round duct is used to make takeoffs from the rectangular plenum or trunk duct to the round duct branch runs (see Figure 8-2). Also, rectangular plenum or trunk ducts or rectangular branch duct takeoffs and runs is used (see Figure 8-3). When used in this manner the takeoff fittings are also completely rectangular.

The space between floor joists is sometimes used as an integral part of the air distribution system. In these cases, the joist space must be made airtight. In some installations, a space above the ceiling or the crawl space beneath the floor may be used as discharge or return air plenums. When these spaces are used for this purpose, they must be sealed airtight, made vapor tight, and insulated. Be sure to check the

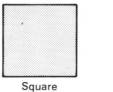

Square

Rectangular

FIGURE 8-1
Square and rectangular duct.

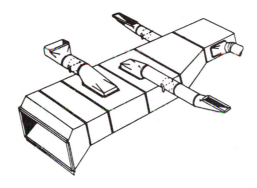

(a) Round pipe takeoff

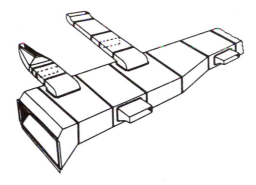

(b) Rectangular pipe takeoff

FIGURE 8-2
Extended plenum round and rectangular duct takeoff connections.

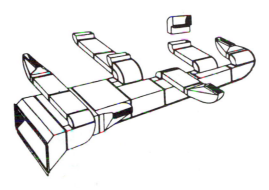

FIGURE 8-3
Rectangular trunk and branch ducts.

local codes and ordinances when using this method. The space between the floor joists may be used as part of the return air system when properly sealed. This space is seldom used for supply air because of the heat transfer through the wall and the possibility of condensation during the cooling cycle.

DUCT MATERIALS

Air ducts are made from a wide variety of materials, such as sheet metal, aluminum, flexible glass fiber, duct board, cement, and tile. Each type of material has certain advantages in a specific application; however, some have equal advantages for the same application and the one used is a matter of preference. Aluminum and glass fiber are fairly light in weight and are less subject to corrosion than sheet metal. However, they both cost more and are more easily damaged. Flexible glass fiber ducts are becoming more common because of the amount of labor saved during installation. Tile and cement pipe are more suitable for installation in a slab floor than the other types because they will not crush from the weight of the concrete used in the slab.

DUCTWORK FITTINGS

Fittings for ductwork are available in almost any shape and any size: adjustable and nonadjustable, round or rectangular in cross section, factory made and shop made are some examples. The fittings illustrated here include elbows, bends, turning vanes, reducers, takeoffs, collars, flexible connections, dampers, end caps, boots, register heads, offsets, floor pans, transitions, and combinations. When a duct system is being designed, careful consideration should be given to the use of fittings that change the direction of airflow or change the duct size. These changes should be kept to a minimum because each change of direction or reduction increases the resistance to airflow. It is not meant that these fittings should not be used but it does mean that their use should be required for the desired system. Thus the duct system should be kept as simple as possible to avoid increasing the resistance unnecessarily. Each one of these fittings add a resistance equal to a given number of feet of straight pipe (see Figures 8-4 through 8-10).

The centerline radii of elbows should be a minimum of one and one-half times the pipe diameter for round ducts. For rectangular ducts the radii should be a minimum of one-half the duct dimension in the turning plane.

All ductwork must be installed permanently, rigid, nonbuckling, and rattle free. All joints should be airtight. Standards 90A and 90B of the National Board of Fire Underwriters specifies the type of materials that may be used in duct manufacture. Generally, supply air plenums and ducts should be constructed of noncombustible materials which are equivalent in strength and durability to those recommended in Table 8-1. The supply air ducts that serve single-family residences are not required to meet these requirements, except for the first 3 ft from the unit. The unit must be a listed unit. They must be con-

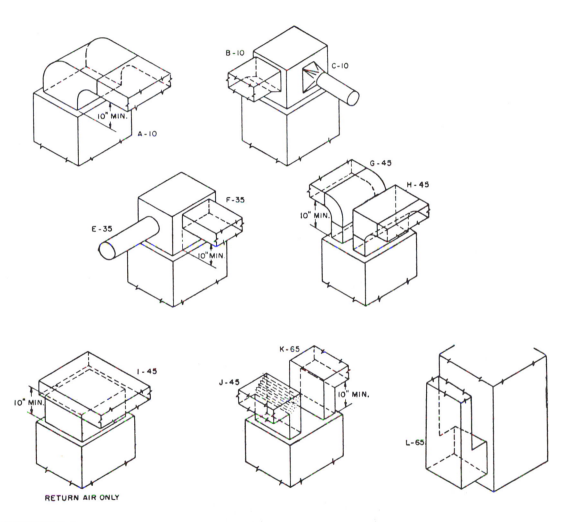

FIGURE 8-4
Equivalent length of supply and return air plenum fittings. (Courtesy of American Society of Heating, Refrigerating and Air-Conditioning Engineers, Inc., Atlanta, Ga.)

TABLE 8-1
Recommended Thickness for Duct Materials

Round Ducts Diameter, In.	Minimum Thickness		Minimum Weight of Tin-Plate
	Galv. Iron, U.S. Gage	Aluminum B&S Gage	
Less than 14	30	26	
14 or more	28	24	IX (135 lb)

Rectangular Ducts Width, In.	Minimum Thickness		Minimum Weight of Tin-Plate
	Galv. Iron, U.S. Gage	Aluminum B&S Gage	
Ducts Enclosed in Partitions			
14 or less	30	26	
Over 14	28	24	IX (135 lb)
Ducts Not Enclosed in Partitions			
Less than 14	28	24	—
14 or more	26	23	—

Note: The table is in accordance with Standard 90B of the National Board of Fire Under-writers. Industry practice is to use heavier gage metals where maximum duct widths exceed 24 in.

(Courtesy of American Society of Heating, Refrigerating and Air-Conditioning Engineers, Inc., Atlanta, Ga.)

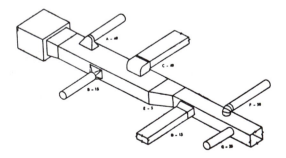

FIGURE 8-5
Equivalent Length of Extended Plenum Fittings.
Add 25 equivalent feet to each of the 3 fittings
nearest the unit in each trunk duct after a reduc-
tion. (Courtesy of American Society of Heating,
Refrigerating and Air-Conditioning Engineers,
Inc., Atlanta, Ga.)

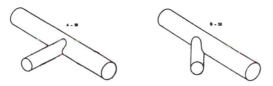

FIGURE 8-6
Equivalent Length of Round Trunk Duct Fittings.
Add 25 equivalent feet to each of the 3 fittings
nearest the unit in each trunk duct. (Courtesy of
American Society of Heating, Refrigerating and
Air-Conditioning Engineers, Inc., Atlanta, Ga.)

structed from a base material of metal or mineral and be properly applied. Also, combustible material that is otherwise suitable for the given application may be used as supply air ducts which are completely encased in concrete.

All supply air ducts should be supported and secured by metal hangers, metal straps, lugs, or brackets. Nails should not be driven through the duct walls and no unnecessary holes should be made in them.

Supply air stacks should not be installed in the outside walls of the building, unless it would be impractical to place them elsewhere.

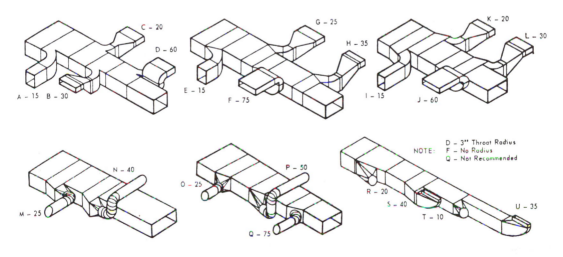

FIGURE 8-7
Equivalent length of reducing trunk duct fittings.
(Courtesy of American Society of Heating, Refrigerating and Air-Conditioning Engineers, Inc., Atlanta, Ga.)

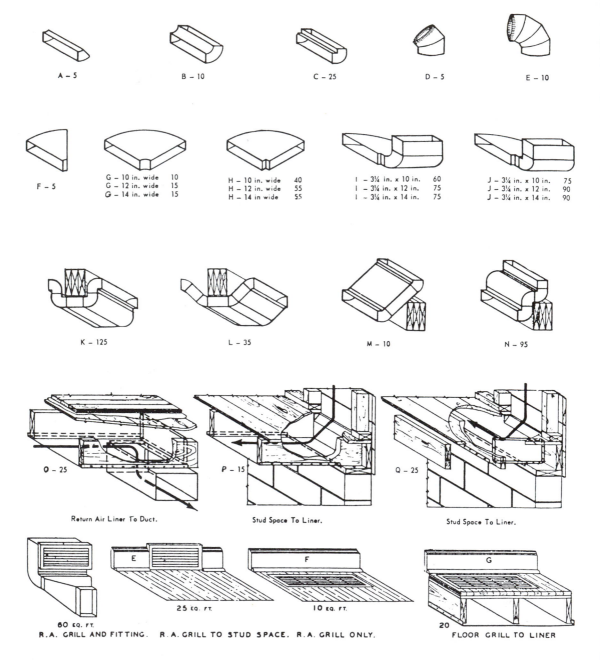

FIGURE 8-8
Equivalent length of angles and elbows for individual and branch ducts. Inside radius for A and B = 3 in., and for F and G = 5 in. (Courtesy of American Society of Heating, Refrigerating and Air-Conditioning Engineers, Inc., Atlanta, Ga.)

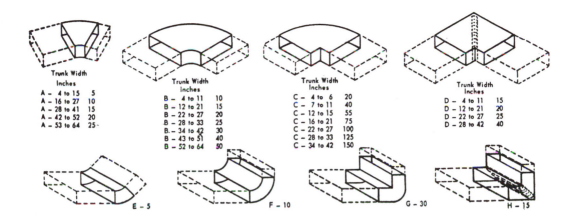

FIGURE 8-9
Equivalent length of angles and elbows for trunk ducts. Inside radius = ½ width of duct. (Courtesy of American Society of Heating, Refrigerating and Air-Conditioning Engineers, Inc., Atlanta, Ga.)

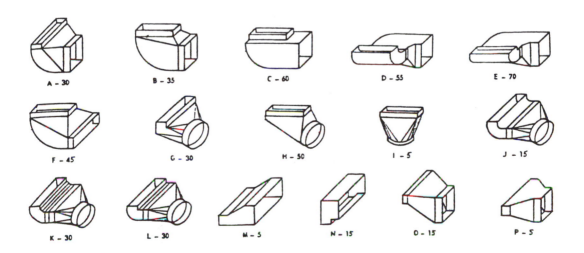

FIGURE 8-10
Equivalent length of boot fittings. These values may also be used for floor diffuser boxes. (Courtesy of American Society of Heating, Refrigerating and Air-Conditioning Engineers, Inc., Atlanta, Ga.)

When necessary to place them in the wall, the stack must be completely insulated against exposure to outside temperatures. Omission of insulation will greatly reduce the heating or cooling capacity of that branch duct, resulting in unsatisfactory conditions in that room.

All supply air ducts should be equipped with an adjustable locking volume damper for air control. This damper should be installed in the branch duct as far from the outlet grill as possible and be completely accessible for adjustment. Return air systems that have more than one return intake may be equipped with volume dampers.

Noise should be eliminated when possible. When metal ducts are used, they should be connected to the unit by pieces of flexible, fire-resistant fabric (flexible connectors). The electrical conduit and piping may increase noise transmission when connected directly to the unit. Return air intakes located next to the unit may also increase noise transmission. Fans located directly beneath a return air grill should be avoided.

INSULATION

When choosing and estimating the insulation to be included in the air distribution system, certain considerations must be given to the code requirements for insulations. ASHRAE Standard 90-75, "Energy Conservation in New Building Design," was developed by the American Society of Heating, Refrigerating and Air-Conditioning Engineers in 1975 "to address new building design for effective utilization of energy." The standard establishes energy-efficient design requirements for:

1. Heating, ventilating, and air conditioning systems and equipment
2. Building exterior envelopes
3. Service water heating systems
4. Electrical distribution systems

These design standards have been adopted by most states as mandatory building codes and standards, either by legislative requirement or executive order. Those states that have not already made this adoption are moving in that direction. A strong push for this acceptance has been given by appropriate federal agencies, which could prevent funding assistance for any new building construction for states that have not established energy codes and regulations.

Listed below are the particular sections of ASHRAE 90-75 that deal with air distribution system insulation:

5.11. Air Handling Duct System Insulation: All ducts, plenums, and enclosures installed in or on buildings shall be thermally insulated as follows:

5.11.1. All duct systems, or portions thereof, shall be insulated to provide a thermal resistance, excluding film resistances, of

$$R = \frac{\Delta T}{15} \text{ ft} - \text{°F/Btu} \quad \text{or} \quad R = \frac{\Delta T}{47.3} \text{ m} - \text{K/W}$$

where ΔT is the design temperature differential between the air in the duct and the surrounding air in °F (K).

Exceptions. Duct insulation is not required in any of the following cases:

a. Where ΔT is 25°F (14 K) or less

b. For supply or return air ducts installed in basements, cellars, or unventilated crawl spaces with insulated walls in one- and two-family dwellings.

c. When the heat gain or loss of the ducts, without insulation, will not increase the energy requirements of the building.

d. Within HVAC equipment

e. For exhaust air ducts

5.11.2. Uninsulated ducts in uninsulated sections of exterior walls and in attics above the insulation might not meet the requirements of this standard.

5.11.3. The required thermal resistances do not consider condensation. Additional insulation with vapor barriers may be required to prevent condensation under some conditions.

5.12. Duct Construction: All ductwork must be constructed and erected in accordance with Chapter 1 of the 1975 *ASHRAE Handbook and Product Directory,* Equipment Volume, or the following NESCA/SMACNA or SMACNA standards:

a. Residential Heating and Air Conditioning Systems—Minimum Installation Standards, August 1973, NESCA/SMACNA.

b. Low Velocity Duct Construction Standards, 4th edition, 1969

c. High Velocity Duct Construction Standards, 2nd edition, 1969

d. Fibrous Glass Duct Construction Standards, 3rd edition, 1972

e. Pressure Sensitive Tape Standards, 1973 (for fibrous glass ducts only).

5.12.1. High-pressure and medium-pressure ducts must be leak-tested in accordance with the applicable SMACNA standard, with the rate of leakage not to exceed the maximum rate specified in that standard.

5.12.2. There is no standard at this time for leak testing of low-pressure ducts. When low-pressure supply air ducts are located outside the conditioned space (except return air plenums), all transverse joints must be sealed using mastic or mastic plus tape. For fibrous glass ductwork, pressure-sensitive tape is acceptable.

5.12.3. There is no standard at this time for damper leakage. Automatic or manual dampers installed for the purpose of shutting off outside air intakes for ventilation air must be designed with tight shutoff characteristics to minimize air leakage.

■ Insulation Types and Uses

There are many types of insulation available for increasing the performance of air distribution systems. The type and thickness required for the particular installation depends on the local and national codes as well as system performance.

The following items explain the various types of insulations available and their uses.

Glass Fiber Duct Wrap. This is a resilient blanket of glass fiber insulation, available unfaced or faced with a reinforced foil kraft vapor barrier facing (FRK) (see Figure 8-11).

It is used to insulate residential and commercial air conditioning or dual-temperature sheet metal ducts operating at temperatures from 40 to 250°F.

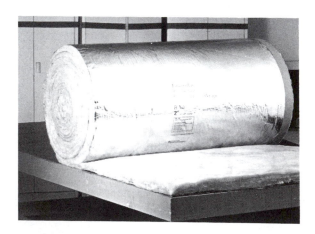

FIGURE 8-11
Fiberglas duct wrap. (Courtesy of Owens-Corning Fiberglas Corp.)

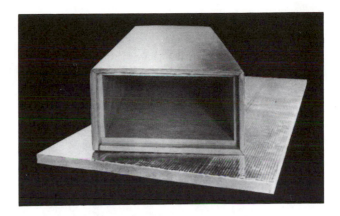

FIGURE 8-12
**Fiberglas duct board. (Courtesy of Owens-Corn-
ing Fiberglas Corp.)**

Glass Fiber Duct Board.   A complete duct product with glass fiber
thermal and acoustical insulation bonded to a tough, flame-resistant
aluminum foil vapor barrier facing, for fabrication of rectangular
ductwork and fittings (see Figure 8-12).

It is used for low-velocity heating, ventilating, and air conditioning duct systems in residential and commercial construction operating at temperatures to 250°F, velocities to 2400 fpm, and 2 in. static
pressure.

FIGURE 8-13
**Duct liner. (Courtesy of Owens-Corning Fiber-
glas Corp.)**

FIGURE 8-14
Duct liner board. (Courtesy of Owens-Corning Fiberglas Corp.)

Flexible Duct Liner.   Duct liner is a bonded mat of glass fiber coated with a black pigmented fire-resistant coating on the side toward the airstream. This coating tightly bonds the surface fibers to resist damage during installation and in service, and provides a uniquely tough airstream surface. It is available in three types, and in thicknesses of ½, 1, 1½, and 2 in. (see Figure 8-13).

Duct liner is designed for use as an acoustical and thermal insulation for sheet metal heating, cooling, and dual-temperature ducts and plenums operating at velocities up to 6000 fpm and temperatures to 250°F. The product is applied to the interior of the ductwork or plenum.

Duct Liner Board.   This is a semirigid bonded board of glass fiber, coated with a flame-resistant coating to resist damage during installation and in service (see Figure 8-14).

It is an acoustical and thermal insulation for sheet metal heating, cooling, and dual-temperature ducts and plenums operating at temperatures to 250°F, and velocities to 6000 fpm. It is applied to the interior of the duct.

Flexible Duct.   Flexible duct is a lightweight flexible duct formed with a resilient inner air barrier, glass fiber insulation, and a reinforced vapor barrier jacket (see Figure 8-15). It is used as an air duct or connector on supply and return air lines in residential, industrial, and commercial heating, ventilating, and air conditioning systems operating at temperatures to 250°F. In addition, it can be used on run-

FIGURE 8-15
Flexible duct. (Courtesy of Owens-Corning Fiber-
glas Corp.)

outs to registers, diffusers, and mixing boxes. The flexible feature allows it to conform to gradual bends necessary when connecting air
ducts to diffusers, or when routing air ducts through spaces with
many obstructions.

Installing Duct Wrap Insulation.   Duct wrap insulation is used to reduce heat loss or gain through sheet metal heating and air conditioning ductwork. Properly insulated sheet metal ducts can mean that
warmer air in the winter and cooler air in the summer will reach the
rooms served by the ductwork, rather than losing or gaining heat energy through the duct walls.

It is important to know that the higher the R value of the insulation, the greater the insulating value. It is equally important to realize
that the R value of duct wrap insulation, once it is installed, depends
on how much it is compressed during installation. If it is compressed
too much, it will lose some of its insulating value.

Full Installed R Value.   The instructions outlined here, if followed
exactly, will assure you of getting the full installed R value of the duct
wrap insulation. The installed R value is printed on the facing of insulation by most manufacturers. It also appears on the product labels
in each package of both faced and unfaced duct wrap insulations.

The installed R value may be obtained when the duct wrap is compressed to no less than 75% of its "as manufactured" thickness after
being installed.

TABLE 8-2

If you are using duct wrap insulation which has these thicknesses printed on label:		Cut the duct wrap insulation to the lengths shown below, depending on shape of duct:		
Nominal (As-manufactured) Thickness (in.)	Average Installed Thickness (in.)	Round and Oval Ducts	Square Ducts	Rectangular Ducts
1.0*	¾	P + 7.0	P + 6.0	P + 5.0
1.5	1⅛	P + 9.5	P + 8.0	P + 7.0
2.0	1½	P + 12.0	P + 10.0	P + 8.0
3.0	2¼	P + 17.0	P + 14.5	P + 11.5
4.0	3	P + 22.0	P + 18.5	P + 14.5

*Available in California only.

(Courtesy of Owens-Corning Fiberglas Corp.)

The key to correct installation is to cut each piece of duct wrap insulation to the stretch-out length listed in Table 8-2. This will prevent the insulation from being overcompressed (see Figure 8-16).

To obtain the full installed R value of insulation, use the following steps:

1. Check the product label to determine the type and thickness of the duct wrap insulation you have bought (see Figure 8-17).

2. Determine the shape and perimeter (P) dimension of the bare sheet metal duct. This dimension will be the circumference of round and oval ducts, or twice the height plus twice the width of square and rectangular ducts. Use Table 8-2 to determine the amount of duct wrap you must cut from the roll (the stretch-out length).

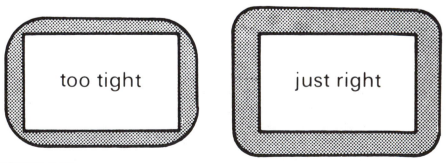

FIGURE 8-16
Proper insulation tightness. (Courtesy of Owens-Corning Fiberglas Corp.)

(A) Wrap type
(B) Installed R-value
(C) Out-of-package R-value, thickness, width, and amount

FIGURE 8-17
Insulation label. (Courtesy of Owens-Corning Fiberglas Corp.)

EXAMPLE: Your sheet metal duct measures 8 in. high and 12 in. wide. The perimeter (*P*) dimension is twice the 8-in. height plus twice the 12-in. width: $2(8) + 2(12) = 16 + 24 = 40$ in. total. If you are using 2-in.-thick duct wrap insulation, you should cut it to $P + 8$ in., or 48 in., to obtain the 1.5-in. average installed thickness and the full installed *R* value for the type of insulation that you bought.

Make the following checks before wrapping the ducts:

1. Check to see that the sheet metal ducts are tightly sealed. Air leakage at sheet metal joints wastes energy and will reduce the effectiveness of any insulation installed on the ducts. If the duct joints are sealed, it will help the duct wrap insulation to do its job.

2. Collect the tools that will be needed. These are:

    a. Large utility shears, or a sharp serrated kitchen knife

    b. Tape measure, or a yardstick and some string

    c. A marking pen (a felt-tip marker is best)

    d. A straightedge (a yardstick will serve here)

    e. 2½- or 3-in. pressure-sensitive tape (foil or vinyl, to match the facing on the duct wrap insulation)

Now you are ready to cut, wrap, and tape the insulation. There are three simple steps to install the duct wrap insulation:

1. Cut:

    • For faced duct wrap insulation:

        a. Use the felt-tip marker and the straightedge to mark a fine

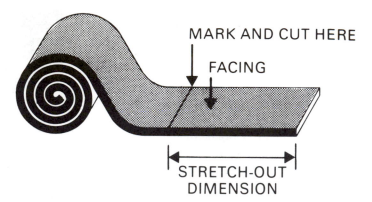

FIGURE 8-18
Measuring insulation. (Courtesy of Owens-Corning Fiberglas Corp.)

line across the facing of the duct wrap for cutting a piece to the correct stretch-out dimension (see Figure 8-18). Cut along this line with the facing up, using the utility shears or the serrated kitchen knife.

b. Cut away a 2-in.-wide piece of the fiberglass insulation, being careful not to cut through the facing. This will provide a 12-in. overlapping tape flair (see Figure 8-19).

• For unfaced duct wrap insulation:

a. Use the felt-tip marker and the straightedge to mark a line across the duct wrap for cutting a piece to the correct stretch-out dimension (see Figure 8-20).

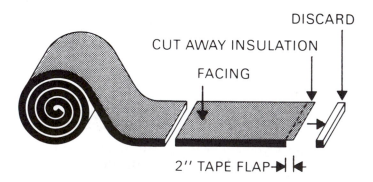

FIGURE 8-19
Cutting insulation. (Courtesy of Owens-Corning Fiberglas Corp.)

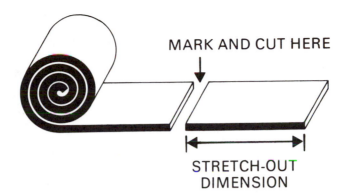

MARK AND CUT HERE

STRETCH-OUT
DIMENSION

FIGURE 8-20
Cutting for unfaced duct wrap. (Courtesy of
Owens-Corning Fiberglas Corp.)

    b. Cut along this line using the utility shears or the serrated
kitchen knife.

2. Wrap:
- For faced duct wrap insulation:

    a. Wrap the insulation around the duct, with the facing outside
and the glass fiber insulation against the duct. The tape flap
should overlap the insulation and the facing at the other end
of the piece of duct wrap. The insulation should be tightly
buttoned (see Figure 8-21).

    b. Next tape the flap with a short piece of tape in the center of
the flap. Be careful not to pull too hard on the tape flap.

- For unfaced duct wrap insulation:

    a. Wrap the insulation around the duct. One end of the duct

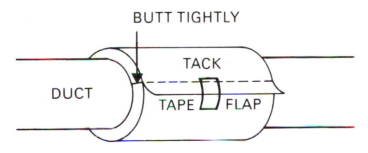

BUTT TIGHTLY

TACK

DUCT

TAPE FLAP

FIGURE 8-21
Wrapping faced duct wrap insulation. (Courtesy
of Owens-Corning Fiberglas Corp.)

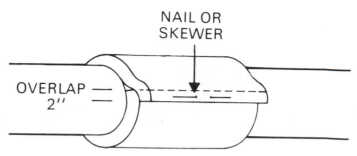

FIGURE 8-22
Wrapping unfaced duct wrap. (Courtesy of Owens-Corning Fiberglas Corp.)

wrap should overlap the other end by about 2 in. (see Figure 8-22). Next, tuck the overlap by stitching the insulation with finishing nails. Do not pull the insulation too tightly around the duct.

3. Tape:

- For faced duct wrap insulation:

    a. First tape the flap seam, using pressure-sensitive foil or vinyl tape compatible with the insulation facing. Rub the tape firmly with your hand. Be careful not to puncture the facing. Patch any holes or tears in the facing with tape (see Figure 8-23).

    b. As you move along the duct, repeat the foregoing steps, butting each piece of duct wrap tightly against the previously installed piece so that the facing flap that runs the length of the roll of wrap is overlapping (see Figure 8-24).

    c. Next, tape all the way around each butt joint, continuing the steps above until the duct is completely wrapped and taped (see Figure 8-25).

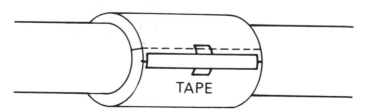

FIGURE 8-23
Taping faced duct wrap. (Courtesy of Owens-Corning Fiberglas Corp.)

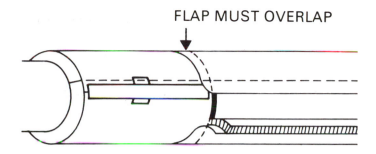

FLAP MUST OVERLAP

FIGURE 8-24
Taping faced duct wrap (continued). (Courtesy of Owens-Corning Fiberglas Corp.)

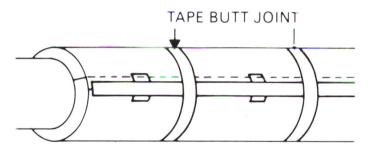

TAPE BUTT JOINT

FIGURE 8-25
Taping all joints of faced duct wrap. (Courtesy of Owens-Corning Fiberglas Corp.)

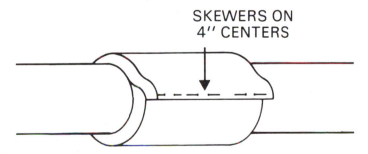

SKEWERS ON 4″ CENTERS

FIGURE 8-26
Securing unfaced duct wrap. (Courtesy of Owens-Corning Fiberglas Corp.)

- For unfaced duct wrap insulation:
  a. Secure the duct wrap insulation overlap by stitching it together using nails or skewers 4 in. apart (see Figure 8-26).
  b. Repeat step a, overlapping 2 in. or butting each piece of duct

BUTT OR OVERLAP TIGHTLY

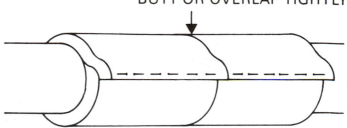

FIGURE 8-27
Joining unwrapped duct wrap. (Courtesy of
Owens-Corning Fiberglas Corp.)

wrap tightly to the next until the duct is completely wrapped (see Figure 8-27).

SUMMARY

Duct systems can be put into two general classifications when used for ventilation: (1) systems where the movement of air is of prime importance without regard to quiet operation or power economy, and (2) systems where air must be moved quietly and with power economy.

Pressure losses in duct systems are caused by the velocity of the airflow, number of fittings, and friction of the air against the sides of the duct.

Round ducts are generally preferred because they can carry more air in less space. Also, less material is used, and therefore there is less surface to reduce the amount of friction and heat transfer through the duct, and less insulation is required than for other popular shapes.

In practice, rectangular duct is used for the plenums and round duct is used for the branch runs and usually the trunk line.

In some installations, a space above the ceiling or the crawl space beneath the floor may be used as discharge or return air plenums. These spaces must be sealed airtight.

Ducts are made from materials such as sheet metal, aluminum, flexible glass fibers, duct board, cement, and tile.

Fittings for ductwork are available in almost any shape and any size: adjustable and nonadjustable, round or rectangular in cross section, factory made and shop made are some examples.

Changes in airflow through a duct should be kept to a minimum because each change of direction or reduction increases the resistance to airflow. Each fitting used adds a resistance equal to a given number of feet of straight pipe.

All ductwork must be installed permanently, rigid, nonbuckling, and rattle free. All joints should be airtight.

All supply air ducts should be supported and secured by metal hangers, metal straps, lugs, or brackets. Nails should not be driven through the duct walls and no unnecessary holes should be made in them.

All supply air ducts should be equipped with an adjustable volume damper for air control.

Return air systems that have more than one return intake may be equipped with volume dampers.

Noise should be eliminated when possible. When metal ducts are used, they should be connected to the unit by pieces of flexible, fire-resistant fabric (flexible connector).

When choosing and estimating the insulation to be included in the air distribution system, certain considerations must be given to the code requirements for insulations.

REVIEW QUESTIONS

1. What is the purpose of ducts in an air distribution system?

2. Where are rectangular ducts generally used?

3. In practice, what type of duct is used for trunk lines and branch runs?

4. Under what circumstances may the space between floor joists be used as part of the return air system?

5. Why are tile and cement pipes more suitable than other types to be installed in a slab floor?

6. For round ducts, what should be the minimum centerline radii of elbows?

7. What organization specifies the type of materials that may be used in duct manufacture?

8. Is it good practice to drive nails through duct walls?

9. Where should the damper be installed in a branch duct?

10. What is the requirement for installing volume dampers in return air ducts?

11. What should be installed between metal ducts and the unit?

12. What is the particular standard that deals with the insulation of air distribution systems?

13. What prevents moisture condensation on the ducts that carry cold air?

14. Should duct wrap insulation be stretched tight around the ducts?

15. Where is duct liner installed?

9 System Cost Estimating

The proper cost estimation of the overall installation for an air conditioning system cannot be overemphasized. This is where the money is either made or lost. Many people claim that there is no break-even point. When no profit is made, the contractor loses money, because he could have been working on another job where he could have made some money.

INTRODUCTION

Care should be taken when estimating the price of an air conditioning unit installation. When the bid is too high, the job will probably be given to another contractor. If the bid is too low, the contractor will lose money. Thus a markup of a required percentage of the materials, equipment, and labor costs plus an appropriate profit should be included in the price to the customer. The price of the job together with a list of what the contractor proposes to do and a list of what is required of others should be presented to the customer. This is sometimes entered on proposal forms, whereas other contractors prefer a typewritten letter for each job. This is a matter of personal preference.

ESTIMATION FORM

A checklist type of estimation form should be used which lists every possible piece of equipment, material, subcontracts, and labor that may be required for the installation (see Exhibit 9-1). Often there are things on the list that will not be used on every installation; however, they do not need to be given consideration in this instance.

The estimate form should include the item, size, type, quantity, cost each, combined cost, percent markup, selling price, and a total. This form need not be shown to the customer; only the proposal is given to the customer.

■ Estimation Example

We will now estimate the cost of the unit that we designed for our model house in previous chapters. (Note: It should be remembered that these prices are for illustration purposes only. The individual making an estimate should use the current prices when the estimate is being made.)

It is usually easier to start at the top of the price estimate form and calculate each item, as needed, on an individual basis.

Step 1. Enter the customer's name, address, date, and estimator as required on the top of the form. This identifies the job, the date, and the estimator for future reference should there be any question about the estimation (see Exhibit 9-2).

Step 2. Enter the size of the condensing unit, the model number, cost, quantity, combined cost, markup, and selling price. When we check the supplier for the cost of our equipment, the price is found to be $431.46. We will assume that a 30% markup will provide us with sufficient money for all our office expenses, insurance, truck costs, labor, and so on, and a fair profit margin. However, the markup should be determined for each individual company by previous experience or other suitable method.

Thus, when the condensing unit cost is multiplied by 30%, the selling price will be $560.90. Enter this figure in the proper column.

Step 3. The cost of the furnace, from the supplier, is $138.51. With a markup of 30%, the selling price is $180.06. Enter these figures in the proper column on the form.

Step 4. The evaporator costs $113.72. Thus a selling price of $147.84 is entered on the form.

Step 5. The thermostat is a T-87 F Honeywell, which requires a sub-base. The cost of the thermostat is $13.52. The subbase costs $10.46. The combined cost is $23.98. The selling price is found to be $31.17. Enter these figures on the form (see Figure 9-1).

Step 6. The subtotals of the equipment can now be determined by adding the columns "Combined Cost" and "Selling Price" and entering the sum of each column in the proper place. The combined cost is $707.67 and the selling price is $919.97.

Step 7. The supply plenum presents another problem. The supplier must be consulted to get the proper price. We will use the enclosed price list for all the metal duct and fittings (see Figure 9-1). Refer to the section of the price list for plenums. There is an example on estimating the cost. Our plenum size is $19\frac{3}{8} \times 22 \times 48$. Thus we have $19\frac{3}{8} + 19\frac{3}{8} + 22 + 22 = 82.75$. We will select 26-gauge metal, and a height of 48 in. shows a multiplier of 0.473. Thus $0.473 \times 82.75 = \$39.14$. The selling price is $50.88.

Step 8. The furnace is to set on a platform in a closet. Thus, the return plenum would consist of the material to make the platform. This is an educated guess in most cases.

Step 9–15. Refer to the price list in Figure 9-1 to calculate the cost and selling price for these steps.

Step 16. The supply grills are found in tables like the one presented in Table 9-1. Locate the cost in the table and enter it in the proper column and calculate the selling price and enter it in the proper place.

Step 17. The prices for return air grills are also placed in tables. In some cases, however, it is better to get the price from your local supplier. In our case they cost $4.77 each, with a selling price of $12.40. Notice that the required opening is 24 in. × 18 in. Since our platform is only 12 in. high, we can divide the return grill into two grills of size 24×9 and put one on another wall.

The remainder of the material section is completed in the same way. The combined cost and selling price are calculated and entered in the subtotal line in their respective places.

The "Subcontract and Labor" section is approached in the same manner. When subcontractors give an estimate, or when the estimator makes an estimate, the figure is entered in the proper column. Some

TABLE 9-1
Grill Price List

Duct Size W × H	List Price	Free Area Sq. Inch	Duct Size W × H	List Price	Free Area Sq. Inch
8 × 6	$2.80	33	16 × 6	$4.30	70
			16 × 8	4.85	96
10 × 4	2.90	27			
10 × 6	3.00	42	18 × 6	4.50	77
10 × 8	3.55	57	18 × 8	4.90	105
12 × 4	3.00	33	20 × 6	5.60	84
12 × 6	3.30	51	20 × 8	6.35	115
12 × 8	3.85	70			
			24 × 6	6.00	103
14 × 4	3.10	39	24 × 8	7.30	140
14 × 6	3.50	61			
14 × 8	4.20	83	30 × 6	7.15	129
			30 × 8	8.85	176

(Courtesy of Standard Perforating & Mfg., Inc.)

companies put a markup on subcontract labor and others do not. Again, this is a matter of preference and depends on the method used for calculating profit. In our example we did not place a markup on any item in this section.

The labor is probably the most difficult item to estimate. Usually, it is much easier when the estimator has worked with the same crew for some time. Thus he can more accurately judge the amount of time required for the job. We estimated that a journeyman installer and a trainee would take 30 hours to install this unit in an existing home. We used the hourly labor cost of $10.00 for the journeyman and $5.00 for the trainee. The installation labor would be $15.00 × 30 = $450.00.

We estimated 1 hour for a service technician to start up and test the unit, for a cost of $45.00. The permits were estimated at $25.00. The local cost will probably be different.

We used the price of $50.00 per ton for warranty reserve, for a total of $175.00. This warranty reserve is to pay for any labor or materials used to maintain the unit through the warranty period. The equipment manufacturer will usually furnish compressors and other electrical equipment for the first year. Few of them, however, will pay any labor charges for replacing these components.

The local tax rate should be used to calculate any taxes due. In our example the rate is 5% on equipment and materials only. Labor, permits, and so on, are not taxable. However, they may be in another area. Be sure to check with the local authorities.

The bid price can now be determined by adding the equipment subtotal, materials subtotal, subcontract and labor, and any miscellaneous charges.

Each of these items is found under the respective heading. The miscellaneous charges are the permits and taxes, for a total of $96.72. The subcontract and labor subtotal is $1020.00. The materials subtotal is $514.52 and the equipment subtotal is $919.97. The sum of the subtotals is $2551.21, the bid price of the job. This is the price that the customer pays.

		3″	4″	5″	6″	7″	8″	9″	10″	12″	14″	16″	18″	20″	22″	24″
DUCT PIPE Snaplock #100 Duct Pipe	30 ga	73.66	74.86	77.89	83.81	99.98	115.58	132.34	146.69	198.65	—	—	—	—	—	—
	28 ga	—	—	—	—	—	—	—	—	—	242.71	271.70	351.38	458.83	—	—
	26 ga	—	—	131.98	158.88	180.86	202.88	229.75	259.09	303.07	383.75	481.52	559.72	620.82	689.26	743.03
	24 ga	—	—	175.98	210.21	241.98	273.75	305.54	337.30	400.84	510.84	—	—	—	—	—
ELBOWS #125 90° Elbows	30 ga	1.27	1.27	1.33	1.49	1.64	1.89	—	—	—	—	—	—	—	—	—
	28 ga	—	—	—	—	—	—	2.53	2.74	3.96	6.82	—	—	—	—	—
	26 ga	—	2.78	2.78	3.05	3.39	3.75	4.57	4.90	7.16	9.18	9.52	13.81	21.28	—	—
	24 ga	—	—	—	4.02	4.34	5.07	6.53	7.97	10.84	13.30	14.17	20.21	25.98	37.56	47.54
#150 45° Elbows	30 ga	—	1.20	1.24	1.35	1.47	1.56	—	—	—	—	—	—	—	—	—
	28 ga	—	—	—	—	—	—	2.19	2.46	3.22	5.73	—	—	—	—	—
	26 ga	—	2.08	2.08	2.29	2.55	2.82	3.41	3.70	5.39	6.88	8.15	11.39	—	—	—
	24 ga	—	—	—	3.22	3.45	4.02	5.20	6.36	8.65	10.63	11.33	16.19	—	—	—
DAMPERS In Line Damper 100D		—	2.74	2.82	3.12	3.39	3.68	4.46	4.63	5.77	7.98	11.75	14.72	—	—	—
Damper Blank 300		—	.32	.32	.40	.50	.57	.67	.76	.80	1.37	2.34	2.97	3.83	4.72	6.76
Damper Blank w/Quadrant 300Q		—	1.30	1.30	1.39	1.64	1.81	2.04	2.29	2.69	3.37	3.96	4.72	6.02	7.94	8.50
Butterfly Damper		—	—	—	—	5.07	—	—	8.74	8.90	—	—	—	—	—	—
Blastgate Damper		—	12.88	—	15.05	16.13	17.18	—	21.49	—	—	—	—	—	—	—
REDUCERS #400 Price based on big end		—	1.94	2.44	2.21	2.32	2.55	2.63	2.80	3.49	4.65	6.27	9.56	12.84	16.97	18.29
STARTING COLLARS Metal Starting Collar 500		.93	.93	.97	1.07	1.20	1.33	1.41	1.77	2.17	2.55	3.43	4.10	6.15	7.03	
Metal Starting Collar with Damper 500D		2.34	2.36	2.59	2.86	3.16	3.41	3.68	4.23	5.20	6.57	10.27	13.66	15.01	16.38	
Tapered Starting Collar with Damper 500TD		—	4.55	4.72	4.97	5.33	6.29	6.63	7.52	9.94	—	—	—	—	—	
Duct Board Starting Collar 500DB		2.21	2.21	2.21	2.50	2.80	3.12	3.39	3.66	4.25	5.16	6.29	7.77	—	—	
Duct Board Starting Collar 500DBD		3.73	3.83	3.83	4.38	4.82	5.43	5.81	6.61	7.70	8.76	10.44	12.44	—	—	
Spin-In Duct Board Starting Collar 500DBS		2.53	2.53	2.53	2.86	3.20	3.54	3.89	4.23	4.95	5.98	7.22	8.88	—	—	
Spin-In Duct Board Starting Collar With Damper 500DBSD		4.02	4.17	4.23	4.80	5.24	5.91	6.29	7.14	8.36	9.52	11.37	13.62	—	—	
45 Side Take-off 510		—	5.18	5.41	5.75	6.00	6.86	8.88	10.55	13.56	—	—	—	—	—	
45 Side Take-off (with Damper) 510D		—	7.28	7.52	8.29	9.01	10.55	12.06	13.54	15.85	—	—	—	—	—	

#500 #500D
#500TD #500DB
#500DBD #500DBS
#500DBSD
No. 100D Inline Damper No. 300Q Damper with Quadrants
510 510D

FIGURE 9-1
Sample duct work price list.

STACK BOOTS

201
Universal Stack Boot

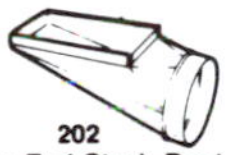

202
Center-End Stack Boot

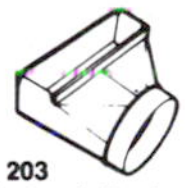

203
90° Angle Stack Boot

	201	202	203
8 x 3¼—5" φ	3.77	3.96	3.96
10 x 3¼—5" φ	4.90	5.66	5.66
10 x 3¼—6" φ	3.87	4.46	4.46
10 x 3¼—7" φ	4.90	5.66	5.66
12 x 3¼—7" φ	3.95	4.63	4.63
14 x 3¼—8" φ	4.63	5.09	5.09
16 x 3¼—9" φ	5.20	6.15	6.15

REGISTER BOOTS

201-R

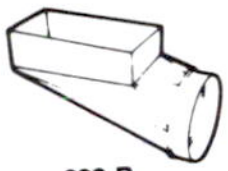

202-R

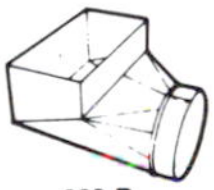

203-R

	201-R	202-R	203-R
8 x 4—5" φ	x	5.73	4.63
10 x 2¼—(5, 6)" φ	3.87	5.75	4.63
10 x 4—(5, 6)" φ	3.87	5.75	4.63
10 x 6—6" φ	x	5.75	4.63
12 x 2¼—(6, 7)" φ	4.10	6.10	5.09
12 x 4—(6, 7)" φ	4.10	6.10	5.09
12 x 6—7" φ	x	6.10	5.09
14 x 2¼—(6, 7, 8)" φ	4.63	6.48	5.75
14 x 4—(6, 7, 8)" φ	4.63	6.48	5.75
14 x 6—8" φ	x	6.48	5.75
14 x 8—9" φ	x	10.46	9.49
16 x 6—9" φ	x	10.46	9.49

STACKHEADS

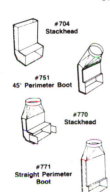

#704 Stackhead

#751 45° Perimeter Boot

#770 Stackhead

#771 Straight Perimeter Boot

	704	751	770	771
8 x 4	4.50	6.76	6.76	6.76
10 x 6	4.67	7.37	7.37	7.37
12 x 6	5.09	8.08	8.08	8.08
14 x 6	6.00	9.66	9.66	9.66
14 x 8	x	10.23	10.23	10.23
16 x 6	7.49	10.23	10.23	10.23
18 x 6	9.37	11.28	11.28	11.28
20 x 6	x	14.17	14.17	14.17
24 x 6	x	16.65	16.65	16.65
30 x 6	x	19.62	19.62	19.62

REGISTER BOXES

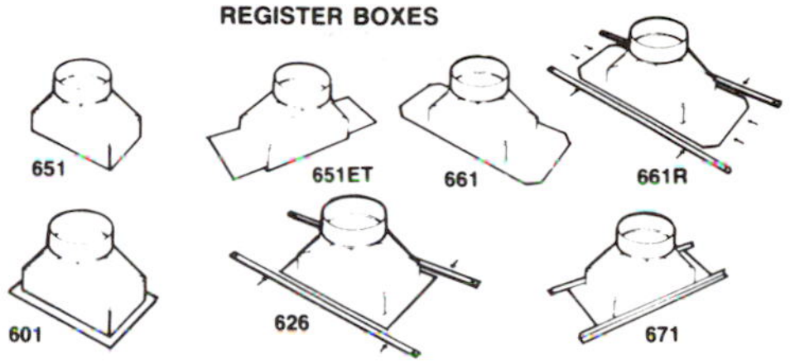

651

651ET

661

661R

601

626

671

Most Popular Sizes

	651	651ET	661	661R	601	626	671
8 x 4—5" φ	2.95	3.22	3.77	4.36	3.49	4.10	4.10
10 x 6—6" φ	3.20	3.43	4.10	4.67	3.66	4.23	4.23
12 x 6—7" φ	3.30	3.54	4.23	4.80	3.87	4.44	4.44
14 x 6—8" φ	3.43	3.68	4.34	4.90	3.96	4.55	4.55
14 x 8—9" φ	4.32	4.61	4.97	5.62	4.97	5.62	5.62
16 x 6—9" φ	4.32	4.61	4.97	5.62	4.97	5.62	5.62

Other Sizes Available

	651	651ET	661	661R	601	626	671
6 x 6—(5, 6)" φ	3.64	3.94	4.76	5.33	4.48	5.07	5.07
8 x 4—(4, 6)" φ	3.92	4.21	5.03	5.60	4.74	5.33	5.33
8 x 6—(5, 6)" φ	3.92	4.21	5.03	5.60	4.74	5.33	5.33
8 x 8—(6, 7, 8)" φ	3.92	4.21	5.03	5.60	4.74	5.33	5.33
10 x 6—(5, 7)" φ	3.92	4.21	5.03	5.60	4.74	5.33	5.33
10 x 8—8" φ	4.29	4.63	5.52	6.08	5.24	5.81	5.81
10 x 10—(6, 7, 8, 9, 10)" φ	4.29	4.63	5.52	6.08	5.24	5.81	5.81
12 x 6—(6, 8") φ	4.29	4.63	5.52	6.08	5.24	5.81	5.81
12 x 8—8" φ	5.54	5.87	6.82	7.41	6.57	7.14	7.14
12 x 12—(8, 9, 10, 12)" φ	5.73	6.06	6.82	7.41	6.57	7.14	7.14
14 x 6—(6,7)" φ	4.55	5.16	5.85	6.46	5.85	6.46	6.46
14 x 8—(8, 10)" φ	7.12	7.45	8.27	8.86	8.27	8.86	8.86
14 x 14—(8, 9, 10, 12)" φ	8.61	9.01	9.56	10.15	10.15	10.72	10.72
16 x 6—(8, 10)" φ	7.12	7.45	8.15	8.72	8.72	9.28	9.28
16 x 8—(8, 9, 10)" φ	7.12	7.45	8.15	8.72	8.72	9.28	9.28
18 x 6—10" φ	7.94	8.23	9.45	10.02	10.02	10.59	10.59
18 x 8—10" φ	8.84	9.14	10.80	11.37	11.37	11.96	11.96
20 x 6—10" φ	8.84	9.14	10.80	11.36	11.37	11.96	11.96
20 x 8—12" φ	9.18	9.87	11.16	11.75	11.75	12.32	12.32
24 x 6—10" φ	11.16	11.83	13.81	x	14.38	x	x
24 x 8—12" φ	14.10	12.48	14.44	x	15.01	x	x

SNAP-ON RAILS .57/pr.

FIGURE 9-1 (Continued)

680

690

680D   690D

Round Ceiling Drop
No Damper

Round Ceiling Drop
with Pull Chain Damper

CEILING DROPS

	680	690	680D	690D
6	4.55	4.55	5.41	5.41
8	4.55	4.55	5.59	5.59
10	5.03	5.03	6.32	6.32
12	6.00	6.00	7.53	7.53
14	6.78	6.78	8.87	8.87

Price based on Large End

INSULATED BOXES

IB

	COLLAR SIZE					
	4" φ	5" φ	6" φ	7" φ	8" φ	9" φ
6 x 6	5.73	5.73	x	x	x	x
8 x 4	5.73	5.73	x	x	x	x
8 x 6	x	5.73	5.73	x	x	x
8 x 8	x	x	6.18	6.18	x	x
10 x 6	x	x	6.18	x	x	x
10 x 10	x	x	7.05	7.05	7.05	x
12 x 6	x	x	6.25	6.25	x	x
12 x 8	x	x	x	x	7.05	x
12 x 12	x	x	x	x	8.41	8.41
14 x 6	x	x	6.82	6.82	6.82	x
14 x 8	x	x	x	x	8.19	8.19
14 x 14	x	x	x	x	x	9.09
16 x 6	x	x	x	x	8.19	8.19
16 x 8	x	x	x	x	10.50	10.50

ROOF JACKS AND FLASHINGS

Roof Jack/Coolie Cap
RJCC

Roof Jack/Banded Cap
RJBC

Stack W/CC

Stack W/BC

Model No.	Diameter	RJCC	RJBC	BC	Stack W/CC	Stack W/BC
# 3	4½"	7.50	11.55	5.55	5.62	9.23
# 4	5½"	8.55	12.48	6.00	6.39	9.98
# 5	6½"	9.59	14.73	6.53	7.21	11.82
# 6	7½"	11.16	16.51	7.28	8.39	13.21
# 7	8½"	14.19	19.39	7.94	9.96	15.48
# 8	9½"	16.42	24.56	11.00	12.32	19.69
#10	10"	19.69	27.19	13.23	—	—
#12	12"	23.50	34.81	19.69	—	—
#14	14"	26.78	47.68	24.94	—	—
#16	16"	35.45	64.60	32.31	—	—
#18	18"	44.00	82.19	41.11	—	—
#20	20"	52.93	99.79	51.36	—	—
#24	24"	81.42	146.79	80.76	—	—
#30	30"	124.73	196.97	118.19	—	—

Less 10% for 12 qty. — one size one pitch
Prices above are for standard pitches (²/₁₂, ³/₁₂, ⁴/₁₂, ⁵/₁₂, ⁶/₁₂ and ⁷/₁₂).
Request quotation for other pitch.

WYES & TEES

#800 WYE

#900 TEE

PRICE FROM
LARGE END

	WYE #800	WYE #800S	TEE #900
3"	—	—	4.77
4"	—	5.87	4.77
5"	4.39	4.82	5.27
6"	4.39	4.82	5.27
7"	4.21	4.98	5.57
8"	4.39	5.07	6.14
9"	4.66	5.32	7.00
10"	5.07	6.27	7.71
12"	6.46	7.37	9.23
14"	8.89	9.75	10.23
16"	11.60	11.60	—
18"	15.28	15.28	—
20"	19.12	19.12	—
22"	24.60	24.60	—
24"	27.31	27.31	—

PLENUMS   PRICE PER INCH OF GIRTH

PLENUM HEIGHT	28GA	26GA	24GA	Add for Duct Liner ½"	1"
to 18"	.209	.236	.289	.209	.236
over 18" to 24"	.264	.289	.341	.264	.289
over 24" to 30"	.316	.341	.421	.316	.341
over 30" to 36"	.341	.368	.446	.341	.368
over 36" to 42"	.393	.421	.498	.393	.421
over 42" to 48"	.421	.473	.578	.421	.473
over 48" to 60"	.525	.578	.682	.525	.578

Quantity Discount
6   LESS 5%
24   LESS 10%

Example: Plenum 16 x 20 x 36   26GA
Girth = 16 + 20 + 16 + 20 = 72
Price = 72 x .368 = 26.50
For ½" D. L. Add 72 x .341 = 24.55

Turbine Flashing

XL Ventilator Cap

BX
Adjustable Pitch
Turbine Flashing

CF
Conduit
Flashing

	TF	XL	BX	CF
3"	x	x	x	4.89
4"	x	x	x	5.43
5"	x	x	x	7.14
6"	11.30	23.51	11.30	9.87
7"	x	24.31	x	x
8"	12.48	24.94	12.48	x
10"	13.80	29.38	13.80	x
12"	11.82	41.11	11.82	x
14"	20.62	70.53	20.62	x
16"	26.53	88.13	x	x
18"	32.31	110.50	x	x
20"	36.76	x	x	x
24"	52.52	x	x	x
30"	78.78	x	x	x

FIGURE 9-1 (Continued)

VENT-A-HOOD DUCT PRODUCTS

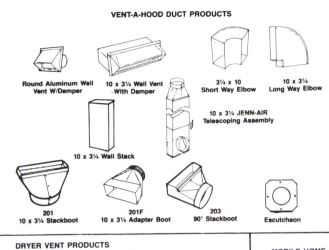

Round Aluminum Wall Vent W/Damper   10 x 3¼ Wall Vent With Damper   3¼ x 10 Short Way Elbow   10 x 3¼ Long Way Elbow

10 x 3¼ JENN-AIR Telescoping Assembly

10 x 3¼ Wall Stack

201 10 x 3¼ Stackboot   201F 10 x 3¼ Adapter Boot   203 90° Stackboot   Escutcheon

PRODUCT	SIZE	PRICE
Round Wall Vent With Damper	5"	10.78
	6"	10.78
	7"	9.34
	8"	13.00
10 x 3¼ Wall Vent	10 x 3¼	9.34
Elbow	3¼ x 10 SW	3.41
	10 x 3¼ LW	6.18
Wall Stack	10 x 3¼ x 30	4.98
	10 x 3¼ x 36	6.03
	10 x 3¼ x 48	7.98
201 Stackboot	10 x 3¼ x 5	4.91
	10 x 3¼ x 6	3.87
	10 x 3¼ x 7	4.91
201F Adapter Boot	10 x 3¼ x 5	5.66
	10 x 3¼ x 7	5.66
203 90° Angle Stackboot	10 x 3¼ x 5	5.66
	10 x 3¼ x 7	5.66
Escutcheon	4" thru 7"	.68
Jenn-Air Telescoping Assembly less boot	10 x 3¼— (5", 6") φ	24.94

DRYER VENT PRODUCTS

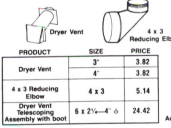

Dryer Vent   4 x 3 Reducing Elbow   6 x 2¼ Dryer Vent Telescoping Assembly W/Boot to 4" φ Adjustable from 66" to 116"

PRODUCT	SIZE	PRICE
Dryer Vent	3"	3.82
	4"	3.82
4 x 3 Reducing Elbow	4 x 3	5.14
Dryer Vent Telescoping Assembly with boot	6 x 2¼—4" φ	24.42

ESS AND DRIVE CLEATS

		Price Per 100 Ft.	
		Cut to Lgth	10 Ft. Lgth
1⅛" Drive Cleat	28GA	26.15	20.90
	26GA	28.76	23.51
	24GA	31.40	26.14
1" Reinforced ESS Cleat	26GA	39.26	34.01
	24GA	44.52	39.27
1⅛" Standing ESS Cleat	26GA	55.02	49.77
	24GA	65.63	60.28
1⅝" Standing ESS Cleat	22GA	83.92	78.67

PITCH PANS AND DRAIN PANS

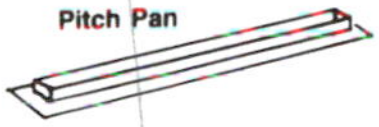

Pitch Pan   Drain Pan

1 Ft.	7.35
2 Ft.	8.03
3 Ft.	9.21
4 Ft.	13.16
5 Ft.	16.48
6 Ft.	19.78
7 Ft.	23.10
8 Ft.	26.44
10 Ft.	33.10

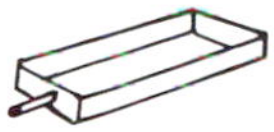

12 x 30	26 ga.	16.03
12 x 36	26 ga.	17.08
12 x 48	26 ga.	18.39
14 x 30	26 ga.	18.39
14 x 36	26 ga.	19.69
24 x 24	26 ga.	19.69
30 x 30	26 ga.	26.26
24 x 24	18 ga.	31.51

MOBILE HOME DUCT KIT PRODUCTS

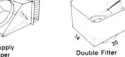

Floor Return Air Grille   AC Supply Damper   Double Filter Return Air Box 14 x 20

12" φ Starter    Butterfly Damper 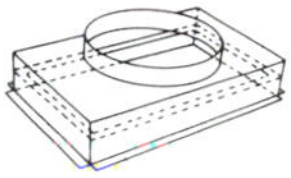   Single Filter Return Air Box

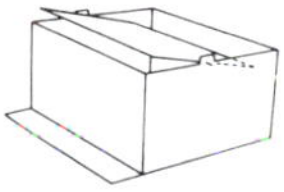

Backdraft Damper

#8451 DUCT KIT

1	14 x 20	DBL Filter Box
2	10 x 20	Filters
1	14 x 20	Floor Grille
1	12" φ	Starter

12" AC KIT

1	12 x 20	Single Filter Box w/12" φ Collar
1	12 x 20	Filter
1	12 x 20	Floor Grille
1	12" AC	Supply Damper

14" AC KIT

1	12 x 20	Single Filter Box w/14" φ Collar
1	12 x 20	Filter
1	12 x 20	Floor Grille
1	14" AC	Supply Damper

MSD KIT

1	14 x 20	Single Filter Box w/14" φ Collar
1	14 x 20	Filter
1	12" φ	Starter
1	14 x 20	Floor Grille

PRODUCT	SIZE	PRICE
Double Filter Box	14 x 20 w/14" φ collar	18.42
Single Filter Box	12 x 20—12" φ	15.12
	12 x 20—14" φ	17.05
	14 x 20—14" φ	17.05
Floor Grille	12 x 20	17.69
	14 x 20	20.55
12" φ Starter		2.55
Butterfly Damper	10" φ	9.44
	12" φ	9.62
AC Damper	12" φ	21.83
	14" φ	22.51
Backdraft Damper		23.42
#8451 Duct Kit		45.47
12" AC Kit		59.12
14" AC Kit		61.39
MSD Kit		42.29

FIGURE 9-1 (Continued)

Customer Name _____ Address _____

Date _____ Estimator _____

EQUIPMENT

ITEM	SIZE	MODEL NUMBER	COST EACH	QUANTITY	COMBINED COST	PERCENT MARKUP	SELLING PRICE
Condensing Unit							
Furnace							
Evaporator							
Humidifier							
Thermostat							
Electronic Filter							
Subtotal							

MATERIAL

ITEM	SIZE	MODEL NUMBER	COST EACH	QUANTITY	COMBINED COST	PERCENT MARKUP	SELLING PRICE
Plenum, Supply							
Plenum, Return							
Round Duct							
Rectangular Duct							
Starting Collars							

EXHIBIT 9-1
Air conditioning price estimating form.

MATERIAL (Continued)							
ITEM	SIZE	MODEL NUMBER	COST EACH	QUANTITY	COMBINED COST	PERCENT MARKUP	SELLING PRICE
Flexible Duct							
Elbows							
Stack Boots							
Register Boots							
Stack Heads							
Wyes							

EXHIBIT 9-1 (Continued)

				MATERIAL (Continued)			
ITEM	SIZE	MODEL NUMBER	COST EACH	QUANTITY	COMBINED COST	PERCENT MARKUP	SELLING PRICE
Tees							
Reducers							
Supply Grills							
Ceiling Diffusers							
Return Air Grill							
Combustion Air Box							
Refrigerant Lines							
Copper Fittings							

EXHIBIT 9-1 (Continued)

MATERIAL (Continued)							
ITEM	SIZE	MODEL NUMBER	COST EACH	QUANTITY	COMBINED COST	PERCENT MARKUP	SELLING PRICE
Suction Line Insulation							
Refrigerant							
Vent, Double Wall							
Elbows, Double Wall							
Vent Conne Connector							
Flashing							
Collar							
Vent Cap							
Drain Line							
Duct Insulation							
Duct Board							
Duct Tape							
Duct Screws							
Duct Hanger							
Control Wire							
Concrete Slab							
Condensate Pump							
Misc. Materials							
Subtotal							

EXHIBIT 9-1 (Continued)

			SUBCONTRACT AND LABOR				
ITEM	SIZE	MODEL NUMBER	COST EACH	QUANTITY	COMBINED COST	PERCENT MARKUP	SELLING PRICE
Freight							
Trucking and Rigging							
Foundations							
Ductwork							
Insulation							
Cutting and Patching							
Painting							
Electric Wiring							
Plumbing							
Oil Tank Installation							
Gas Connection							
Labor							
Startup and Test							
Permits							
Cleaning of Existing Ducts							
Service Reserve							
Taxes							
Misc.							
Subtotal							

BID PRICE	
Equipment Subtotal	_____
Materials Subtotal	_____
Subcontract and Labor	_____
Miscellaneous Charges	_____
TOTAL	_____

EXHIBIT 9-1 (Concluded)

Customer Name	Model House				Address	Dallas, Texas	

Date __10–5–82__   Estimator _____ Langley _____

EQUIPMENT

ITEM	SIZE	MODEL NUMBER	COST EACH	QUANTITY	COMBINED COST	PERCENT MARKUP	SELLING PRICE
Condensing Unit	3.5 tons	CTD-351-QC	431.46	1	431.46	30	560.90
Furnace	100,000 Btu	CF 100-BD-3	138.51	1	138.51	30	180.06
Evaporator	—	UCC-41-23	113.72	1	113.72	30	147.84
Humidifier							
Thermostat		T-87F Honeywell	13.52/10.46	1	23.98	30	31.17
Electronic Filter							
Subtotal					707.67		919.97

MATERIAL

ITEM	SIZE	MODEL NUMBER	COST EACH	QUANTITY	COMBINED COST	PERCENT MARKUP	SELLING PRICE
Plenum, Supply	19⅜ × 22 × 48	—	39.14	1	39.14	30	50.88
Plenum, Return	Platform		20.00 Estimated	1	20.00	30	26.00
Round Duct	10″		1.47	15ft.	22.00	30	28.60
	7″		1.00	25 ft.	24.99	30	32.49
	8″		1.16	25 ft.	28.90	30	37.57
	9″		1.32	15 ft.	19.85	30	25.81
	6″		0.84	15 ft.	12.57	30	16.34
Rectangular Duct							
Starting Collars	10″		1.41	2	2.82	30	3.67
	9″		1.33	1	1.33	30	1.73
	7″		1.07	1	1.07	30	1.39

EXHIBIT 9-2
Air conditioning price estimating form.

MATERIAL (Continued)							
ITEM	SIZE	MODEL NUMBER	COST EACH	QUANTITY	COMBINED COST	PERCENT MARKUP	SELLING PRICE
Flexible Duct							
Elbows	7″	Adj.	1.64	4	6.56	30	8.53
	10″	Adj.	2.74	2	5.48	30	7.12
	8″	Adj.	1.89	1	1.89	30	2.46
	9″	Adj.	2.53	2	5.06	30	6.50
	6″	Adj.	1.49	2	3.98	30	5.17
Stack Boots							
Register Boots	12 × 3¼ × 70	201	3.95	3	11.85	30	15.41
	8 × 4 × 60	601	4.74	2	9.48	30	12.32
	14 × 6 × 80	601	3.96	1	3.96	30	5.15
	12 × 6 × 80	601	5.24	1	5.24	30	6.81
Stack Heads							
Wyes	10-7-8	# 800	5.07	2	10.14	30	13.18
	8-6-6	# 800	4.39	1	4.39	30	5.70

EXHIBIT 9-2 (Continued)

ITEM	SIZE	MODEL NUMBER	COST EACH	QUANTITY	COMBINED COST	PERCENT MARKUP	SELLING PRICE
			MATERIAL (Continued)				
Tees							
Reducers	9 × 8	# 400	2.63	1	2.63	30	3.42
Supply Grills	8 × 4	# 351	2.80	2	5.60	30	7.28
	12 × 4	# 351	3.00	3	9.00	30	11.70
	12 × 6	# 351	3.30	2	6.60	30	8.58
Ceiling Diffusers							
Return Air Grill	24 × 9		4.77	2	9.54	30	12.40
Combustion Air Box	200 sq. in. (5″ × 40″)		18.81	1	18.81	30	24.45
Refrigerant Lines	3/8″		{ 91.50 set	1	{ 91.50 set	30	118.95
	3/4″		{	1	{		
Copper Fittings							

ITEM	SIZE	MODEL NUMBER	COST EACH	QUANTITY	COMBINED COST	PERCENT MARKUP	SELLING PRICE
Suction Line Insulation	Included in line set						
Refrigerant	Included in line set						
Vent, Double Wall	5″		12.67	2 (10′)	25.34	30	32.94
Elbows, Double Wall							
Vent Connector	5″		1.87	1	1.87	30	2.43
Flashing	5″		5.95	1	5.95	30	7.74
Collar	5″		1.53	1	1.53	30	1.99
Vent Cap	5″		6.63	1	6.63	30	8.62
Drain Line	¾″	copper pipe	0.72	8′	5.76	30	7.45
	¾″ × 90° Ell	copper	0.65	1	0.65	30	0.85
Duct Insulation	2″		77.95	1 roll	77.95	30	101.34
Duct Board							
Duct Tape	2″		5.32	1 roll	5.32	30	6.92
Duct Screws	⅜″ × ½″		1.86	1 lb.	1.86	30	2.42
Duct Hangar	1 roll		7.34	1 roll	7.34	30	9.54
Control Wire	18-2/18-4		0.03/0.06	25′/15′	1.62	30	2.11
Concrete Slab	26 × 32 × 4		32.50	1	32.50	30	42.25
Condensate Pump							
Misc. Materials	coil housing	HUH-36-48-23	37.82	1	37.82	30	50.34
	Dampers 7″/6″/9″/8″		3.39/3.12/ 4.46/3.68	3/2/1/1	24.55	30	31.92
Subtotal					394.90		514.52

MATERIAL (Continued)

EXHIBIT 9-2 (Continued)

SUBCONTRACT AND LABOR							
ITEM	SIZE	MODEL NUMBER	COST EACH	QUANTITY	COMBINED COST	PERCENT MARKUP	SELLING PRICE
Freight							
Trucking And Rigging							
Foundations							
Ductwork							
Insulation							
Cutting And Patching							
Painting							
Electric Wiring	Line voltage to both units		275.00	1	275.00		275.00
Plumbing							
Oil Tank Installation							
Gas Connection	Gas line to furnace		75.00	1	75.00		75.00
Labor			450.00	15.00 × 30 hrs.	450.00		450.00
Start Up And Test			45.00	1 hr.	45.00		45.00
Permits			25.00		25.00		25.00
Cleaning of Existing Ducts							
Service Reserve			175.00	50.00 per ton	175.00		175.00
Taxes					1434.49	0.05	71.72
Misc.							
Subtotal							1116.72

BID PRICE	
Equipment Subtotal	919.97
Materials Subtotal	514.52
Subcontract and Labor	1020.00
Miscellaneous Charges	96.72
TOTAL	2551.21

Index